Department of Health

GUIDELINES FOR THE SAFE PRODUCTION OF HEAT PRESERVED FOODS

London: HMSO

CONTENTS

1 FOREWORD

These Guidelines replace the Food Hygiene Code of Practice No.10, "The Canning of Low Acid Foods", issued in 1981 under Section 13(8) of the Food and Drugs Act 1955. That Act has since been replaced by the Food Safety Act 1990.

The 1981 Code required updating in the light of legislative changes and technological developments in canning processes. This publication is designed to provide a guide to those concerned with the production of heat preserved foods, importers and distributors of canned goods, educational establishments, food producers, and Environmental Health Officers. It does not purport to be comprehensive or to render legal advice. The law is as stated as at June 1994.

The Guidelines are consistent with hazard analysis and critical control point (HACCP) principles now being widely promoted internationally.

Thanks are due to the members of the working party at Campden Food and Drink Research Association, which was set up to revise the 1981 Code, and to the various companies and organisations which took part in the subsequent consultation exercise.

2 INTRODUCTION AND SCOPE OF DOCUMENT

2.1   These guidelines are concerned with the safe manufacture
       of foods which are made commercially sterile principally
       by the application of a predetermined amount of lethal
       heat and which are packaged in any type of
       hermetically sealed container in order to prevent
       microbiological infection after commercial sterilisation.

       The scope of this document is therefore wider than the
       earlier (1981) Code of Practice No. 10, The Canning of
       Low Acid Foods. It includes all ambient stable foods,
       generally recognised as canned foods, including
       aseptically processed foods and acidified (normally low
       acid) foods. It excludes, however, cured meat products
       which require chilled storage, products such as jams and
       pickles which depend primarily for their stability on
       chemical preservation, and naturally acidic canned
       products. Milk is also specifically excluded as specific
       regulations exist.

       Of particular importance to the success of the safe
       production of heat preserved foods are steps to ensure:

       a. The prevention of pre-processing spoilage or
          contamination;
       b. The maintenance of container integrity;
       c. The establishment and consistent application of the
          scheduled process by properly trained, experienced and
          supervised personnel;
       d. The prevention of post-processing contamination.

       The effectiveness of good manufacturing practice (GMP)
       therefore depends on application of established
       technological principles within an organised framework,
       concepts which are embodied in total quality
       management (TQM) which includes quality control,
       assurance and audit.

2.2   The Guidelines do not set out to be an exhaustive treatise
       on canning or aseptic processing, but are intended to
       provide advice and guidance to those concerned with the

production of heat preserved foods. Each manufacturer should be able to demonstrate an effective system of quality management appropriate to the individual circumstances, which implement the underlying principles of these Guidelines. The preparation of the Guidelines have had the benefit of expert assistance and advice of technologists from the United Kingdom industry.

3 GLOSSARY OF TERMS

3.1 **Acidified food**
A food, any component of which has a neutral pH, to which acidic ingredient(s) are added to bring the pH of all components to 4.5 or below.

3.2 **Ambient stable**
Commercially sterile under the intended temperature conditions of storage. Refrigerated storage is not required.

3.3 **Aseptic process**
A process in which the product and container are separately sterilised and then combined together under commercially sterile conditions.

3.4 **Bleed**
A small diameter, permanently open vent on the upper surfaces or instrument pocket of a steam retort.

3.5 **Blown can**
A can which is distended or burst due to the production of gas, by microbiological or chemical means, within the can.

3.6 **Bringing up**
The process by which a retort or steriliser is heated up to its scheduled sterilising temperature.

3.7 **BS 5750/ISO 9000**
A British/international standard for quality management of all aspects of a business.

3.8 **Bulk incubation**
The incubation of substantial amounts of product in order to establish the incidence of blown cans.

3.9 **Burst test**
A test in which compressed air is introduced, in a controlled manner, into a semi-rigid or flexible container until structural failure results. The reading obtained provides an indication of seal strength for welded seals.

3.10   **Canned**
Product in containers which have been hermetically sealed and heated to achieve commercial sterility.

3.11   **Cap**
The closure for a glass jar. Hermetic sealing is obtained between the lining compound of the cap and the upper rim of the jar.

3.12   **Chlorination**
The process for microbial disinfection by exposure to an aqueous solution of a chlorine-containing substance. The principal materials used are chlorine gas, chlorine dioxide or sodium hypochlorite.

3.13   **Clean in place (CIP)**
The process by which a continuous flow plant is cleaned by successive wash and rinse treatments without dismantling the plant.

3.14   *Clostridium botulinum*
The most heat resistant of the pathogenic organisms. Clostridium botulinum is a spore-forming mesophilic anaerobe. It produces a lethal toxin which attacks the central nervous system of man.

3.15   **Coliform organisms**
The coliform group includes all Gram-negative, non-sporeforming rods capable of fermenting lactose with the production of acid and gas at 37°C in less than 48 hours (WHO International Standards for Drinking Water).

3.16   **Commercial sterility (appertization) of food**
The condition achieved by application of heat which renders food free from viable micro-organisms, including those of known public health significance, capable of growing in the food at the temperatures at which the food is likely to be held during distribution and storage.

3.17   **Condensate**
The water produced by the condensation of steam.

3.18   **Conduction heating**
The process of heating in which heat energy is transferred by vibration between adjacent molecules but without bulk movement of material.

3.19 Cook
The stages of the sterilising operation during which cans are heated up and then held at sterilising temperatures.

3.20 Cook room
The area of a factory in which the thermal processing equipment is situated.

3.21 Cook temperature
The temperature of the retort specified for the sterilising operation.

3.22 Cook time
The length of time during which the sealed containers are totally exposed to the specified cook temperature.

3.23 Crate
The perforated vessel used to hold containers inside a retort during the sterilising process.

3.24 Critical factor
A factor specified in the scheduled thermal process, an alteration in the value of which may cause a change to the sterilisation value achieved by that process.

3.25 Cured meat
A meat or meat product partially preserved by the addition of curing salt.

3.26 Curing salt
Sodium nitrite, which is used for its inhibitory effect on the outgrowth of microbial spores in cured canned meat products.

3.27 Disinfection
The application of effective chemical or physical agents or processes to a cleaned surface or the application of an effective chemical or physical agent to a water supply with the intention of reducing the numbers of micro-organisms to a level at which they can be reasonably assumed to present no risk to health.

3.28 Double seam
The means by which a can end and can body are fixed together to form an hermetic seal. The end curl and body

flange are interlocked and then tightened in two
mechanical rolling operations.

3.29 **Dud detector**
An on-line device or instrument for detecting containers
which are deficient in respect of a specified parameter
(normally internal vacuum).

3.30 **Filling temperature**
The temperature of each component of the food at the
time of filling into the container.

3.31 **Flat sour**
Micro-organisms which, as a result of their metabolism,
do not evolve gas but which reduce the pH of the food
substrate. Spoilage is therefore not accompanied by
swelling of the container.

3.32 **F_0 value**
A measure of the amount of lethal heat which results
from a specified thermal process in the slowest heating
part of the can. The number is the lethal effect
equivalent to the number of minutes at 121.1°C (250°F)
when assuming instantaneous heating and cooling and a z
value of 10 C degrees (18 F degrees).

3.33 **Good manufacturing practice (GMP)**
The means by which products are made to the specified
quality, in the most economical manner, and with regard
to the safety of both the manufacturing personnel and the
consuming public.

3.34 **Hazard analysis critical control point (HACCP)**
HACCP is a systematic approach to the identification
and assessment of hazards and risks associated with all
the stages of a food operation and the definition of means
for their control.

3.35 **Hazard**
The potential to cause harm. Hazards can be
microbiological, chemical or physical.

3.36 **Headspace**
The volume enclosed between the top surface of the
product within a container and the inner surface of the
container lid or end.

3.37 Heat penetration test
A test in which the temperature at the slowest heating point within the container is monitored during thermal processing in order to calculate the sterilising value achieved.

3.38 Heat seal
An hermetic seal formed by the fusing together of two plastic surfaces by the application of heat.

3.39 Hermetic seal
A seal of a container which is sufficiently tight to prevent the transmission of micro-organisms across the seal.

3.40 High acid food
Any food product in which all components have a pH value of 4.5 or below.

3.41 Holding section
A part of a continuous-flow sterilising plant in which product is maintained for a specific time at an elevated temperature (after the heating section) in order to achieve uniform sterility.

3.42 Incubation test
A test in which containers are held for a suitable time at a temperature specific for the growth of micro-organisms under investigation in order to ascertain whether microbial growth occurs.

3.43 Initial temperature
The temperature of the contents of a container to be processed at the time the sterilising operation begins. The minimum initial temperature to be experienced under manufacturing conditions should be specified in the scheduled process.

3.44 Layer pad
A sheet of perforated material used to separate successive layers of containers within a retort crate. The perforations must be sufficiently large and numerous so that heat transfer from the heating medium during sterilisation is negligibly restricted.

3.45   **Lethal heat**
        The effect of exposure to temperature, under specified
        conditions, transformed mathematically in order to give a
        measure of sterilisation achieved.

3.46   **Low acid food**
        Any food product in which any one or more components
        has a pH value greater than 4.5 at the end of the
        thermal process.

3.47   **Lux**
        A measure of the intensity of light.

3.48   **Mesophilic organism**
        A micro-organism having an optimum growth temperature
        in the region of 35–37°C.

3.49   **Non-Newtonian fluid**
        A fluid in which viscosity varies with rate of shear.

3.50   **Open top container**
        A container supplied with one end through which the
        container is filled with food. The term is usually applied
        to metal cans.

3.51   **Overpressure**
        Pressure applied to a retorting system in excess of the
        pressure of saturated steam at the system temperature.

3.52   **Panelling**
        Permanent mechanical deformation of the walls of a
        container due to a net external applied pressure.

3.53   **Pathogen**
        A micro-organism capable of giving rise to foodborne
        disease in man.

3.54   **Peaking**
        Permanent mechanical deformation of the end(s) of a
        container due to the effect of net internal pressure

3.55   **pH value**
        A measurement of acidity or alkalinity on a scale from 14
        (alkaline) to 1 (acid). It is numerically equal to the
        negative logarithm to base 10 of the hydrogen ion
        concentration.

3.56 **Post-process contamination**
The contamination of a food product in an hermetically sealed container by the ingress of micro-organisms after completion of the thermal process.

3.57 **Post-process operations**
Those operations involving heat sterilised containers which take place subsequent to the heat sterilising operation.

3.58 **Potable steam**
Steam of a suitable quality for incorporation into a product as an ingredient by direct condensation.

3.59 **Potable water**
Water of such chemical and bacterial quality that it is wholesome and fit for human consumption. Potable water is further defined in the EC drinking water directive.

3.60 **Quarantine**
A method of segregation of material pending investigation and subsequent disposal in a controlled manner.

3.61 **Recall procedure**
A documented emergency procedure used to prevent further consumption of product in a suspect batch and for the return of all such material to the supplying company, or suitable alternative location, where it may be fully investigated.

3.62 **Recirculated water**
Water used for cooling purposes which, for conservation reasons, is itself cooled and stored before re-use. The process of recirculation greatly increases the likelihood of microbial contamination and such water should be adequately disinfected before re-use.

3.63 **Residual free chlorine**
The amount of chlorine left in a free and uncombined state in a water supply after any reaction between chlorine and organic matter in the water has occured.

3.64 **Retort**
A pressure vessel designed for the heat processing of food packed in hermetically sealed containers by an

appropriate heating medium and, where necessary, with superimposed air pressure.

3.65   **Retort bleed**
A small orifice or petcock 3–6 mm (⅛"–¼") diameter, through which steam escapes throughout the whole of the heat process.

3.66   **Risk**
An estimate of the probability of a hazard occuring.

3.67   **Saturated steam**
Steam in equilibrium with water at a given pressure (pressure and temperature are dependent upon each other).

3.68   **Scheduled process**
The heat process chosen by the processor, validated for a given product and container size, to achieve commercial sterility.

3.69   **Seal integrity**
The adequacy of a seal which ensures hermeticity.

3.70   **Soil**
Any undesirable material, including food residues, atmospheric dirt, dust etc, which should be removed by cleaning.

3.71   **Steam spreader**
The perforated pipe in a sterilising vessel through which steam issues. The design promotes rapid attainment of even temperature distribution.

3.72   **Thermal process**
The heat treatment given during the sterilisation operation, expressed minimally as a combination of time and temperature.

3.73   **Thermophilic organisms**
Micro-organisms with optimum growth temperatures in the region of 55–60°C. The spores of certain thermophilic organisms are particularly heat resistant.

3.74   **Vacuum**
The **pressure, below atmospheric pressure,** when air has been eliminated from an hermetically sealed container; usually expressed in centimetres or inches of mercury.

3.75   **Venting**
The flushing of air out of steam retorts prior to the commencement of the sterilisation process by means of admission of steam into the retort while the vent orifices or valves remain open.

3.76   **Venting schedule**
A specified procedure for the venting of retorts to ensure adequate removal of air which has been established by a series of temperature distribution tests.

3.77   **z value**
The numerical value obtained by measuring the number of degrees Fahrenheit or Celsius required for the thermal death curve of the test organism in a specific substrate to traverse one log cycle, i.e. the temperature change required to effect a tenfold change in the rate of microbial destruction.

4 MANAGEMENT RESPONSIBILITIES IN THE MANUFACTURE OF HEAT PRESERVED FOODS

4.1   INTRODUCTION

This section is especially addressed to senior management who are responsible for the decision to produce heat preserved foods. Whether or not they are involved in the detailed implementation of their decision, senior managers remain responsible for providing an effective system with adequate resources to ensure the safe production of these foods. They are also responsible for knowing on a continuing basis that their plans remain effective in achieving safe food production through a process of audit and review.

What follows in this section is an overview of the strategy that senior managers should understand and adopt to enable them to produce safe food.

4.2   CONTROL FOR PRODUCT SAFETY

4.2.1   The safety of heat preserved foods is assured by good design and subsequent control during the production, storage and distribution of such products. Finished product testing may be useful but cannot provide a comparable or economic level of assurance, and is therefore less effective as a means of ensuring product safety.

4.2.2   A comprehensive production system should be designed, documented, implemented, and controlled. Additionally, the system should be designed to prevent unsafe product, as a result of operating deviations and failures, from reaching the market place. Appropriate resources must be provided to operate the system.

4.2.3   BS 5750/ISO 9000 are a series of internationally recognised standards for quality systems. Conformity to these standards requires that all company policies and procedures which affect product quality are fully documented. Actual performance is then systematically audited against the documentation in order to check that the system achieves the required levels of product quality.

It is a further requirement of the BS 5750/ISO 9000 standards that the quality system is itself periodically reviewed to ensure that it is still relevant to the ongoing business.

4.3   HACCP/CRITICAL FACTORS

4.3.1   In the safe manufacture of heat sterilised foods, the planning and design stages are fundamental to success. All new projects rely on the application of the wide experience and knowledge of professionals and specialists. It is important, however, to recognise that new projects will embody unique elements, and it is therefore important from the outset to include a rigorous analysis of anticipated hazards.

4.3.2   HACCP is the acronym for hazard analysis critical control point. It is an analytical tool that enables management to introduce and maintain a cost-effective, ongoing food safety programme. HACCP in manufacturing involves the systematic assessment, by a multidisciplinary team, of all the steps that are critical to the safety of the product. It equally may be used both for new and existing product lines. A HACCP study allows management to concentrate technical resource into those manufacturing steps that critically affect product safety. A list of critical control points (CCPs) is produced, together with operating targets, monitoring procedures, and corrective actions for each CCP. For continuing safety, full records must be kept of each analysis, and the efficacy of the study must be verified on a regular basis and when aspects of the operation change.

4.3.3   HACCP is applicable to the identification of microbiological, chemical and physical hazards affecting product safety. The technique can also be used to identify hazards and CCPs associated with microbial spoilage and quality of products. HACCP must be applied to a specific process product combination, either to an existing process or as part of a development brief, and will require the full commitment of senior management and technical staff to provide the resources necessary for successful analysis and subsequent implementation.

4.3.4   One of the many advantages of the HACCP concept is that it will enable a food manufacturing company to move away from a philosophy of control based primarily on end product testing (i.e. testing for failure) to a preventative approach whereby potential hazards are identified and controlled in the manufacturing environment (i.e. prevention of product failure).

4.3.5   HACCP is a logical and cost-effective basis for better decision making with respect to product safety. It provides food manufacturers with a greater security of control over product safety than is possible with traditional end product testing and, when correctly implemented, may be used as part of a defence of "due diligence". HACCP has both national and international recognition as the most cost-effective means of controlling foodborne disease and is endorsed as such by the Joint FAO/WHO Codex Alimentarius Commission.

4.3.6   A practical guide to the operation of HACCP is described in detail in the Campden Food and Drink Research Association's (CFDRA) Technical Manual No. 38.

4.4   AUDIT

4.4.1   Audit is a further management technique in which the effectiveness of factory systems is objectively measured against standards. It is likely that a company will be subject to audit initiated by its customers, but equally the management of any company should be concerned to use internal audit as a means of monitoring its own performance and achieving improvement when necessary. It may also wish to audit its suppliers.

4.4.2   A useful definition of audit is given in BS 4778/ISO 8402(1986): "An audit is a systematic and independent examination to determine whether quality activities and related results comply with planned arrangements and whether these arrangements are implemented effectively and are suitable to achieve objectives."

4.4.3   It is important that the requirement of independence of auditors is fully accepted. This means not only that auditors are free from pressures that may bias a report, but also that they are not so familiar with the subject that bias results from this very familiarity. It also means that auditors should be independent from the auditee for analysis of data and that as individuals they must not be responsible for detailing or progressing corrective or remedial action which they then re-audit.

5 MANUFACTURING PREMISES

5.1   LOCATION

In the United Kingdom there is a requirement under the Food Safety Act 1990 for all food manufacturing premises to be registered with their local Environmental Health Department.

Premises should be sited with due regard for the operations to be carried out in them, for the provision of services needed, and to avoid contamination from and of adjacent activities.

5.2   GENERAL CONSTRUCTION

5.2.1   The structure of the premises should be conducive to good sanitation and should be suitable for the type of food processing carried out. In particular, the premises should provide adequate working space for the various operations performed so that overcrowding of equipment, which may have a deleterious effect on cleaning and standards of hygiene, does not occur. The production lines should be easily accessible from all sides to permit inspection, maintenance and cleaning of equipment.

5.2.2   Walls, partitions and doors should be constructed of materials which provide a durable, smooth, waterproof finish, capable of being easily and thoroughly cleaned. The angles between the walls and floors should be coved, and windowsills, if present, should be sloped towards the floor.

5.2.3   Floors in food processing areas should be hard surfaced, non-absorbent and adequately drained. They should be durable, waterproof, non-toxic in use, and easy to clean and disinfect. They should be slip resistant and without crevices, and should slope evenly and sufficiently for liquids to drain to trapped outlets. Where outlets are fitted with grilles, these should be removable. If the floors are ribbed or grooved to facilitate traction, any grooving of this nature should always run towards a drainage channel. The junctions between the floors and walls should be waterproof and, if practicable, should be coved or rounded for ease of cleaning. Concrete, if not properly finished, is porous and can be affected by animal oils, strong brines, various detergents and disinfectants. If used, it should be dense, of good quality, with a well finished waterproof surface.

5.2.4   The buildings used for processing operations and all ancillary buildings within the site should be maintained in good repair.

5.2.5   All areas around the building and within the boundaries of the site should be maintained in a clean and tidy condition. In particular, discarded equipment and accumulations of processing wastes should not be permitted to remain.

5.2.6   The use of exposed wood in the construction of the premises is undesirable in processing areas. It should be limited to those purposes where there is no satisfactory impervious substitute and, where used, should be treated with sealant or other materials to render it non-absorbent.

5.3 SEPARATION OF PROCESSES

5.3.1 Premises should provide separate physical areas for the various storage and process operations, whilst ensuring good product flow from raw materials through to the storage of finished products. Correct separation of processes will reduce the risk of contamination and spoilage by chemical, foreign body, microbiological or other factors.

Separate areas should be provided for the following purposes:

a. Storage of stocks of unused empty cans or other containers;
b. Storage of ingredients used in the make-up of products;
c. Storage and preparation of raw meat, of poultry and of fish (see current EC regulations);
d. Storage and preparation of vegetables or other materials;
e. The manufacture, filling, closure and sterilisation of the foods;
f. Storage of freshly processed cans until sufficiently cooled and dry;
g. Storage of packaging materials such as cartons, wrapping materials etc;
h. Accumulation of waste materials from the food preparation processes;
i. Storage of by-product materials;
j. Storage of the final finished goods;
k. Lockable storage for cleaning and other chemicals;
l. Engineers' store and workshop;
m. Labelling, cartoning and palletising operations.

5.3.2 It is very important to ensure that unprocessed and processed containers are so separated that under no circumstances can unprocessed containers become mixed with, or mistaken for, those which have been processed.

5.4 SEPARATION OF PERSONNEL

The separation of personnel may provide an essential part of the separation of processes with regard to avoidance of microbiological cross-contamination, as is the case for pre- and post-process container handling operations (see also subsection 6.11). Access to processing areas in which operations of high microbiological risk are being undertaken should be limited to essential personnel.

5.5 LIGHTING

Adequate natural or artificial lighting should be provided throughout the premises. Light bulbs and fluorescent tubes should always be protected whenever breakage could give rise to any form of direct or indirect hazard. A high level of lighting is conducive to good hygiene, and the correct intensity and colour of lighting should be particularly provided for all inspection operations.

5.6 VENTILATION

Adequate ventilation must be provided to prevent build-up of excessive heat and humidity and to minimize condensation on the structure, equipment and products throughout the premises. Mechanically forced ventilation is frequently desirable or necessary. Special attention should be paid to the removal of steam from those areas containing equipment, such as open food boiling pans, and retort areas where considerable steam will be emitted. Account should be taken of contamination arising from other industrial operations in the vicinity. Ventilation openings should be screened for insects, and the screens should be made easily removable for cleaning. Professional advice should be sought on particular ventilation requirements.

5.7 DRAINAGE

5.7.1 A drainage system of suitable construction for the effluent in question, adequate in size and capable of operating efficiently at the maximum load imposed by the processes, including cleaning, at the plant should be provided. Flow of effluent should always be away from high risk areas. Different effluent types, e.g. process effluent and domestic waste, may need to be kept separate. Liquid domestic wastes from toilets must not be able to contaminate food materials or plant.

5.7.2 Drain inlets within the premises should be properly trapped, easily accessible, and covered with suitable removable grids where necessary. Rain water should not be piped into factory drains.

5.7.3   All drains should be designed to prevent back-flushing into the factory in the event of blockage of the sewerage system.

5.7.4   Channel drains provided within the building should be of sufficient size to permit proper cleaning and should be provided with removable grids. Such channels in process areas should be cleaned daily and maintained in a sound condition.

5.7.5   The drainage system within the factory should be regularly cleaned and maintained to prevent the occurrence of blockage and insanitary conditions.

5.7.6   Disposal of plant effluent must meet requirements of the appropriate authorities and advice should always be sought from them.

5.8   HIGH SPECIFICATION AREAS

5.8.1   It is important that the contamination of vulnerable raw materials should be avoided during the preparation processes. Any raw material which is susceptible to increased microbiological contamination, such as raw meat, fish or poultry, or which can cross-contaminate other raw materials should be considered as a vulnerable raw material. Areas used for such preparation should therefore be designed to a high standard to permit cleaning and disinfection processes to be carried out easily and effectively to prevent the build-up of contamination. Separate rooms should be provided for the preparation of specific raw materials, e.g. the cutting and boning of raw meat. A separate facility is required for removing the outer packaging from all raw materials. Specifically, raw meat intended for processing, which is received packed in cartons, must be removed from the cartons in a completely separate area. Discarded cartons must be removed from the unpacking area at an appropriate frequency.

5.8.2   Particular attention should be paid to such high specification rooms, which should be constructed with walls lined with ceramic tiles, stainless steel sheeting or any other metallic or non-metallic lining which will give a smooth, easily cleaned and essentially non-absorbent surface.

5.8.3   Floors should be constructed to a high standard and should be of a durable, slip-resistant, impervious material, coved at the wall-floor junctions.

5.8.4   Ceilings should be properly underdrawn and surfaced with a light coloured, non-absorbent lining capable of being readily cleaned.

5.8.5   Walls and ceilings should, as far as possible, be free from projections. Equipment and services (e.g. electricity, water, steam) should be installed to a safe standard and in such a manner as to avoid pest infestation or accumulation of dirt and to permit access for cleaning.

5.8.6   Lighting should be not less than 540 lux (50 foot candles) and should not affect the appreciation of food colour.

5.8.7   The room should be temperature controlled in accordance with its operational purpose and the requirements of relevant legislation.

5.8.8   Separate hand-washing facilities, in accordance with current EC legislation, should be provided within preparation rooms.

5.9   PREVENTION AND ERADICATION OF PESTS

5.9.1   The most effective contribution towards infestation control is in maintaining good housekeeping standards, both within the factory buildings and the external surroundings; for example, controlling accumulations of food and paper debris, keeping gangways, passages and roadways clear, removing redundant equipment and materials from manufacturing areas, and practising good stock rotation.

5.9.2   Rooms used for the storage and preparation of foodstuffs and packaging materials should be protected against the entry of rodents, birds and insects. Doorways should be protected and windows which open for ventilation purposes should be screened. The screens should be easily removable for cleaning and should be made from corrosion-resistant material.

5.9.3   Ultraviolet electronic devices provide a measure for elimination of flying insects but should be deployed in accordance with makers' instructions and in a manner such that dead insects do not fall into the food processing area. They should be inspected and cleaned regularly. Shatterproof tubes are now available for these devices.

5.9.4   In addition to good housekeeping, a precautionary anti-infestation programme may be advisable but regular recorded inspections by trained personnel should be made throughout the premises, including the areas around the buildings, for evidence of possible infestation. Records should be critically reviewed.

5.9.5   In the event of infestation occurring, appropriate steps should be taken immediately to eradicate the pest concerned. This should be undertaken by experienced and skilled personnel. In the event of re-infestation, a review of the system should be undertaken and appropriate modifications made.

5.9.6   It is essential that the utmost care is taken in the use of poisons or pesticides in food premises to prevent contamination of the food product. All stocks of rodenticides, insecticides and similar toxic substances must be in locked storage away from other materials and only used by suitably trained personnel.

5.9.7   Animals should not be taken into or allowed to enter food handling, raw material or finished product storage areas.

5.10   SANITARY ACCOMMODATION

5.10.1   Separate sanitary accommodation for each sex must be provided to at least the minimum standards required by legislation.

5.10.2   All sanitary accommodation should be constructed to satisfactory standards of hygiene, with smooth, impervious internal surfaces, and fitted with equipment capable of being kept clean and sanitary.

5.10.3   All sanitary accommodation must be adequately lit and must be kept clean.

5.10.4   Rooms containing water closets and urinals must be properly ventilated and must not communicate directly with food rooms or food processing areas. All doors leading to sanitary accommodation should be self-closing.

5.10.5   Adequate wash-hand basins must be provided adjacent to the toilets. They must be provided with hot and cold water or warm water only at a suitably controlled temperature. An appropriate hand-cleaning preparation should be provided, together with single-use towels or other hygienic means of hand drying. It is desirable that taps should be foot, knee or electronically operated (non-hand operable taps are a requirement of some EC directives). Additional hand washing facilities should be available on entry to high risk areas within the factory, and so placed as to ensure that persons entering the food process area have no alternative but to wash their hands, including the microbiological laboratory. (See also subsection 6.5.)

5.10.6   Notices must be displayed in the sanitary accommodation instructing operatives to wash their hands after using the accommodation, as required by current EC legislation.

5.11   STAFF AMENITIES

5.11.1   Adequate separate accommodation for the changing and the storage of outdoor clothing should be provided, preferably adjoining the sanitary accommodation. (See subsection 6.4.5.)

5.11.2   Suitable facilities should be provided for meals. The consumption of food in any production areas of the factory should not be permitted, other than for quality assurance purposes. Areas where smoking is permitted should also be controlled (see subsection 6.5.5).

5.12   QUALITY OF GENERAL WATER SUPPLY

5.12.1   The provision and maintenance of a satisfactory water supply of potable quality in a cannery has important public health implications. It is essential that all water used in the making up of products or likely to come into contact with the product or packaging is at least of the highest standard for drinking water laid down in the EC directive relating to the quality of water for human consumption. Mains water arriving at the plant should be of that standard and the responsibility for ensuring this rests with the water authority.

5.12.2   It should be recognised that water of sound potable quality may be unsuitable as an ingredient in certain products. Water used should be chemically tested at appropriate intervals, both in respect of public health and of product considerations.

5.12.3 In the canning plant, water in the distribution system should be sampled from appropriate outlets. EC standards require that coliforms are not detectable in 100 ml of 95% of samples taken nor in any two consecutive samples of that series. In most plants the total aerobic colony count should be less than 100 organisms per millilitre after incubation for 3 days at 20-22°C. However, in practice it is important that limits should be established for the water in the plant and that any significant variation from these should be investigated immediately. Coliform counts should be performed at least once a month at different points in the plant and total aerobic colony counts at least once a week.

5.12.4 If water is derived from a private well or is further treated on arrival at the premises, samples should also be taken as the water enters the distribution system. All of these should be free from coliforms in 100 ml. Water from a private source should be sampled for coliforms at least once a month at the point of entering the distribution system.

5.12.5 If the water is chlorinated in the plant, residual free chlorine estimations should be carried out, at least twice a day. Once an adequate dosing regime has been established, the frequency of microbiological testing may be reduced.

5.12.6 For container cooling water requirements, see subsection 17.3.

5.13 QUALITY OF GENERAL STEAM SUPPLY

5.13.1 This subsection covers steam for general service, i.e. for process and heating, where steam is not in direct contact with food product. (For Potable Steam, see subsection 5.14.)

5.13.2 Dry saturated steam should be used. It should be generated by the boiler, which is operated and maintained in accordance both with current local regulatory standards or practices and the manufacturer's recommendations.

5.13.3   Superheated steam should generally be avoided due to the reduced coefficient of heat transfer of the "dry gas". However, up to three (3) degrees Celsius of superheat is often acceptable. Typically, this amount of superheating could occur in practice through the throttling effects of pressure/temperature control valves fitted on the steam supply. Some superheat may also be used to prevent undue condensation in an extensive distribution system.

5.13.4   Steam should be dry, clean and free of non-condensables. In order to achieve these requirements, it will be necessary to:

a. Operate the boiler at maximum permissible pressure and avoid peak loads.
b. Provide the appropriate quantity of feed water treatment and maintain ideal total dissolved solids (TDS) levels through properly operated controls. Specialists should be consulted.
c. Install strainers to protect the final heat exchange equipment, control valves, steam traps, etc.
d. Incorporate automatic air vents throughout the pipework distribution system up to the retort to minimize the presence of air and other non-condensables at the point of use.
e. Limit steam pipeline velocities to below 25 m/s.

5.13.5   Steam generators, or central boiler plant, which may be prone to wet steam caused by carry-over, should have steam separators installed near the steam take-off. Separators should be drained using float-type steam traps.

5.13.6   Audible/visual alarm(s) and pressure gauges should be fitted to the steam header serving the retort(s) to confirm the supply or activate an alarm.

5.13.7   Steam supplies may be metered to establish process usage and facilitate costing of the service to the product.

5.13.8   Condensate should be removed from the pipelines, process plant and heat exchangers through properly sized and selected steam traps. Thermodynamic traps are recommended for mains drainage, and float or inverted bucket traps are ideally suited to drain process plant. Where the process permits, condensate should be returned to the boiler feed tank.

5.14   POTABLE STEAM

5.14.1   This subsection covers potable steam, for culinary use, which is in direct contact with food product or food contact surfaces and is used as heating medium and ingredient in the final product.

5.14.2   Clean, dry saturated steam is paramount and it is necessary to make increased provision in respect of these factors. The requirements under subsection 5.13 apply but with the following essential specific or additional stages.

5.14.3   Feed water treatment chemicals must be compatible with direct food use to prevent contamination through taint, odour or any other means which could make it unsafe or unsuitable for human consumption.

5.14.4   Final steam filters must be used which are capable of removing all particles down to 5 micron in size.

5.14.5   All materials of construction, including gaskets and seals, must be compatible with the potable steam, the food process, descaling or cleaning solutions, and product packaging.

5.14.6   In general, stainless steel will be the preferred contact metal. Compatible rubbers and plastic may also be used. In instances where the feed water is distilled/deionised water or is specially treated such that the clean steam will be particularly corrosive, then grade 316L or equivalent stainless steel metal contact parts will be essential.

5.14.7   Equipment must be suitably rated for pressure and temperature and the surface finish, design and construction should be hygienically designed. Equipment should be self-raining.

5.14.8   Condensate from the injected steam is lost but can be recovered and returned to the feed tank from other sources where the process permits. Stainless steel sanitary steam traps are available from major manufacturers.

5.14.9   Dry steam is particularly important in order to avoid excessive condensate mixing with the product. Carry-over of boiler feed water and pipeline condensate should be avoided. A steam separator should be fitted, and a further separator is desirable near the injection point.

5.14.10   A final steam filter, as described in subsection 5.14.4, should be fitted prior to the steam injection point and this should be preceded by a line-size strainer in order to improve life and performance of the filter element.

5.14.11   The above should be regarded as a minimum. Correct installation, operation, maintenance and monitoring should be employed to ensure adherence.

5.15   STEAM BOILER PLANT

5.15.1   The boiler plant should be operated at its maximum permissible pressure, and pressure reducing valves should be used where systems demand lower pressures. It is good practice to install line-sized steam separators after the steam takeoff ore crown valve. These separators should be drained using float-type steam traps.

5.15.2   An appropriate feed water treatment regime should be instigated, with particular attention paid to non-food-compatible agents such as hydrazine. These should NOT be used. The level of TDS in the boiler water is a very important consideration and thought should be given to the fitting of fully automatic blowdown TDS control systems. Acceptable TDS levels are typically in the region of 2,500–3,000 ppm, but the boiler manufacturer and chemical fe'J water treatment specialists should be consulted for blowdown vessel information. It is usual practice to install boiler water sample coolers for safety and also to ensure that samples taken are representative of the boiler water.

5.15.3   Boiler blowdown vessels should be used in preference to blowdown pits. Separate blowdown lines should be taken from the bottom blowdown, boiler TDS control, and gauge glass/level control drains to the blowdown vessel. The installation and subsequent operation of the system should be in accordance with the Health and Safety Executive Guidance Note PM60 (1987).

5.15.4   Boiler plant should be operated and supervised as suggested in the Health and Safety Executive Guidance Note PM5 (revised December 1989). Particular attention should be paid to the testing of automatic level controls and gauge glasses. Where traditional automatic level controls are installed, a trained boiler attendant should be on site AT ALL TIMES when the boiler(s) is in operation. Where boilers are run during quiet periods, it is not always necessary to have a trained boiler attendant on site at all times, but there must be someone on site who is competent to respond to alarms and take appropriate action. In this case, the boiler water level controls MUST be of the HIGH INTEGRITY SELF-MONITORING type. In ALL cases, boiler-controls should be tested on a daily basis by a trained boiler attendant.

5.15.5   Bottom blowdown valves should be operated every day in short sharp bursts to remove sludge, and it should NOT be possible to blowdown more than one boiler at a time. This can be controlled by there being only one blowdown valve key in the boiler house in the case of manually operated valves or, alternatively, interlocking automatic timed bottom blowdown valves.

5.15.6   Boiler feed tanks should be heated using direct steam injection to a temperature in the region of 90–95°C to keep the levels of dissolved oxygen to a minimum. The actual temperature limit will be a function of the net positive suction head (NPSH) of the feed pump with respect to the actual head of the feed water (height of feed tank above the centre of the feed pump suction). Small proprietary mechanical de-aerators are available which can be installed on feed tanks, thereby reducing the amount of chemical oxygen scavengers which need to be used.

5.16   OTHER SERVICES

All air supplies which come into direct contact with product must be fitted with adequate water and oil traps to ensure a clean supply. Similar considerations may apply to other services such as vacuum and controlled atmosphere gases.

6 PERSONNEL

6.1   LEGISLATIVE OBLIGATIONS OF PERSONNEL INVOLVED IN THE MANUFACTURE OF HEAT PRESERVED FOODS

6.1.1 General
The Food Safety Act 1990, the Public Health (Control of Disease) Act 1984, the Public Health (Infectious Diseases) Regulations 1988 as amended, and the Food Hygiene (General) Regulations 1970 as amended (there are separate regulations in Scotland and Northern Ireland) set out certain legal requirements concerning the prevention and spread of food poisoning and the personal cleanliness and hygiene of food-handling personnel. The legal responsibility for complying with these Regulations is shared by management and operatives. The management of a cannery should therefore take steps (including the display of notices, and written notification, where necessary) to ensure that all food-handling personnel are aware of their legal obligations.

6.1.2 Management should be aware of both general and product-specific hygiene requirements imposed by regulations and directives of the European Community.

6.2 MANAGEMENT STRUCTURE

6.2.1 Any organisation involved in the manufacture of canned foods should have a defined management structure in which the authorities and responsibilities for the various aspects of the operation are clearly assigned. In particular, the management of key departments affecting the safety or quality of the foods produced, such as manufacturing or quality assurance, should be by suitably competent persons with full understanding of the operations, including hazards associated with failure of the operations under their control.

6.2.2 Departmental managers normally report to their line manager and by chain of command to the chief executive of the company. It is his ultimate responsibility to ensure that the factory's operations are carried out in accordance with stated company policy.

6.2.3 Designated nominees should be identified to deputise during the absence of key personnel. This line of responsibility may not always be down the chain of command but may, in fact, be upwards to a more senior person in the absence of the key person.

6.2.4 The managers of key areas must have an adequate number and calibre of supporting staff to enable their responsibilities to be satisfactorily undertaken.

6.2.5 All managers from the most senior down should be conversant with basic technical principles of good cannery practice and have a positive attitude towards such principles.

6.3 COMPANY TRAINING POLICY

6.3.1 The company should have a written policy on the training of personnel. This should reflect the need to train personnel to the standard necessary for proper undertaking of their duties.

6.3.2 Training should be given at recruitment and augmented and revised as appropriate. Personnel should understand and comply with standards set out in that training.

6.3.3 Training records should be kept for each individual.

6.3.4   Training should be split into two categories: general training and specific training.

6.3.5   **General Training**
Food businesses should ensure that personnel are supervised, instructed and/or trained in food hygiene matters commensurate with their work activity. Written or verbal examinations or other tests may be used to assess competence. Regular retraining should be undertaken to ensure continued understanding and compliance.

6.3.6   **Specific Training**
Appropriate specific training should be given to manufacturing and support personnel working in specific areas, including product preparation (section 8), filling and closure operations (section 9) and thermal processing operations (sections 10–17). Training for such areas would include principles of canning, cooling water quality control, container seam evaluation, and post-thermal process can handling. This should be augmented by a considerable amount of practical training. Specialised training can be obtained from appropriate institutes such as research associations.

6.3.7   Contractors and temporary staff should receive general training and specific training as necessary. They should understand and abide by the standards set out in the training. Signed acceptance of the training would encourage this. Failure to comply with previously agreed good manufacturing practices may provide reason for dismissal of the contractor from site.

6.3.8   Due regard should be given to the ability of personnel to understand and comply with written instructions, procedures, notices etc. Attention should also be given to overcoming language and reading difficulties.

6.4   HYGIENE STANDARDS

6.4.1   **Legislation**
See subsection 6.1.

6.4.2   Prominent notices should be displayed describing employees' basic legal obligations. In addition, it is good practice to display notices containing specific hygiene instructions. Issue of such notices should be controlled under a quality system procedure to ensure that only accurate and up-to-date information is displayed. On commencement of employment, employees should be made aware of both their legal obligations, in writing, and the management's application of hygiene policy.

6.4.3   The management should ensure that all food handlers have appropriate training in food hygiene, and should provide continuing education in food hygiene for all staff. All employees should be reminded periodically of their responsibilities in personal hygiene and behaviour. A company-issued handbook describing hygienic practices within the operational area helps promote compliance with GMP.

6.4.4   In order to promote an hygienic environment and to encourage compliance with hygiene standards, it is important that the management lead by example. They should be punctilious regarding their own personal hygiene, keep strictly within the site rules, and demand high standards throughout the cannery.

6.4.5   (See also section 5.) Adequate facilities to enable Personal hygiene standards to be met should be made available for use by all employees. This includes toilets for each sex, hand washing and showering facilities, and storage lockers for personal belongings and outdoor clothing. Separate equipment washing areas and facilities should also be provided.

6.4.6   Procedures must be written which describe the frequency and methods of cleaning and disinfection of manufacturing lines. It is especially important that post-process can handling equipment is included. Compliance with and the efficacy of the written procedures must be assessed at regular intervals by independent audit.

6.4.7   Good housekeeping is essential to the hygienic operation of a cannery; consequently, a "tidy as you go" policy should be enforced.

6.5   PERSONAL HYGIENE

6.5.1   In the interest of best hygienic practices, personnel should adopt good personal hygienic standards, thus minimizing the risk of contaminating the food. Appropriate facilities must therefore be provided.

6.5.2   Personnel should thoroughly wash their hands after using toilets or urinals, and also before commencing work at the start of the day and on returning to the production areas after leaving for any reason whatsoever. Preferably, non-perfumed bactericidal soap should be used. Non-hand operated taps, to prevent recontamination of clean hands, should be installed at all sinks. Disposable hand towels or hot air driers should be used to avoid further contamination. Finger nails should be kept short and clean.

6.5.3   Protective overclothing should be worn. It should be clean, washable or disposable, in good repair, and it should be changed when soiled. (For more detail see subsection 6.7.)

6.5.4   All cuts and abrasions on any exposed part of the operative should be covered by a clean waterproof dressing. It is recommended that such dressing be distinctively coloured to assist recovery should it become detached from the food handler.

6.5.5   Smoking or any use of tobacco or other smoking mixtures should be forbidden in all food handling and production areas. If smoking is permitted at all, separate suitably designated areas should be made available (see subsection 6.5.2).

6.5.6   The wearing of jewellery, watches and other adornments (except wedding rings and possibly sleeper earrings) should be prohibited amongst all personnel entering food production areas. The use of make-up should be discouraged and the wearing of false eyelashes, false nails and nail varnish should be prohibited.

6.5.7   Eating and spitting in all food handling and production areas should be forbidden. Drinking water should be obtained from specific foot/elbow operated drinking fountains.

6.6   HEALTH CONTROL OF FOOD HANDLERS

6.6.1   Prior to commencing employment, personnel should complete a detailed medical questionnaire to provide information regarding previous illnesses which may have a bearing on their fitness to work with food. Such information should be passed to an appropriately trained health care worker and cases of doubt as to the interpretation of the questionnaire referred to a medical practitioner. (EC directives may have somewhat varying requirements for medical examination and the relevant legislation should be consulted.)

6.6.2   Food handlers who are suffering from, or are shown to be carriers of, or have been in contact with someone who has an infection which is likely to be transmitted by food should be excluded from work connected with handling food until they are pronounced fit for work by a competent medical officer.

6.6.3   The detection of infected food handlers depends to a very large extent on their co-operation in reporting illness to management who can then obtain medical advice. The EC Directive on the Hygiene of Foodstuffs (93/43/EEC) requires that no person, known or suspected to be suffering from, or to be a carrier of, a disease likely to be transmitted through food or while afflicted, for example with infected wounds, skin infections, sores or with diarrhoea, shall be permitted to work in any food handling area in any capacity in which there is any likelihood of directly or indirectly contaminating food with pathogenic micro-organisms.

6.6.4   Periodic checks should be made by a responsible person to ensure that personnel are free of any septic wounds or cuts on the hands or exposed parts of the body, and that any uninfected wounds or skin lesions are properly protected by waterproof dressings. This is especially important for workers in the post-processing area.

6.7   PROTECTIVE CLOTHING

6.7.1   Protective clothing should be provided, suitable for the tasks undertaken, and should be washable or disposable, clean, in good repair, and changed and/or laundered regularly.

To remove the risk of contaminating the food by buttons, clothing should be fastened with secure metal studs or velcro-type fastening, and pockets should be avoided on the outside of clothing to prevent objects falling into food.

6.7.2   Protective clothing should be worn by all personnel in manufacturing areas. Such clothing is designed to protect the food from contamination by the operators and to protect the wearer's own clothing. The protective clothing should preferably be light coloured, and if segregation or easy identification of personnel is required, colour coding may provide a suitable means. Engineers and other personnel who normally work outside food manufacturing areas may wear dark coloured clothing, but if they are required to work inside the food manufacturing halls, then ideally they also should wear light coloured clothing.

6.7.3   Coats should cover the arms and may be of the loose-fitting smock type to aid comfort.

6.7.4   Headgear should completely contain the hair or an additional hairnet should be worn. If personnel have beards, then beard masks are required. It must be stressed that headgear is NOT a fashion item and therefore must be worn correctly.

6.7.5   Trousers ideally should have an elasticated waist and a velcro-type fly.

6.7.6   Footwear should be kept clean and hygienic, and should be of an appropriate safety standard for the working environment.

6.7.7   If there is a need to indicate segregation of high-risk production areas from low-risk production areas, for example in post-process can handling areas, this can be helped by high-risk personnel wearing clothing of a different colour to the rest of manufacturing.

6.7.8   Protective clothing should not be worn outside the relevant manufacturing area.

6.8   TECHNICAL SERVICE

6.8.1   The company should have, or have access to, a suitably qualified department capable of providing necessary technical support to the factory.

6.8.2   The technical department must be capable of establishing, measuring and recording any parameters of ingredients and processes necessary to produce the correct finished product, including facilities to meet legally enforceable standards; for process establishment and re-establishment; for monitoring line hygiene; to establish sensory properties of finished product; and for equipment calibration.

6.9 PERSONNEL DIRECTLY CONCERNED WITH STERILISING OPERATIONS

6.9.1   Staff working in this area must be competent, responsible and diligent (see subsection 16.1).

6.9.2   Specific training should be given and should include all relevant aspects of sterilising processes required in order to undertake the particular job satisfactorily (see subsection 16.1).

6.9.3   Staff should receive appropriate training before assuming responsibilities in sterilising areas and their competency should be confirmed by management at regular intervals.

6.10 PERSONNEL INVOLVED IN POST-PROCESS HANDLING

6.10.1   High standards of personal hygiene must be maintained by personnel working in such areas as it is possible to contaminate processed cans. (See also subsection 6.6.4.).

6.10.2   Training MUST be given to post-process can handling personnel to ensure that they understand the need for line hygiene, product segregation and destruction of uncontrolled containers, and the control of goods which require segregation as identified by a quarantine note, and that they do NOT handle wet cans.

6.11 RESTRICTION OF PERSONNEL BETWEEN FACTORY AREAS

6.11.1   Movement of personnel between the various factory areas should be controlled to minimize any likelihood of contamination of either food or containers. Production areas ideally should not be used as throughfares.

6.11.2   Identification of personnel in different areas may be achieved by colour coded clothing.

6.11.3   Certain personnel may be exempt from access restrictions between areas. These may include maintenance, hygiene and technical support, QA auditors and management. However, if such personnel move between high and low risk areas, they must be required to change their protective clothing.

6.11.4   Personnel requiring access to restricted areas should obtain relevant authorisation before entering the restricted areas.

6.11.5   It should be a company policy to control site access of non-employees and restrict, for security reasons, their access to manufacturing areas.

7 INGREDIENT RAW MATERIALS

7.1   SPECIFICATIONS

7.1.1   Written specifications should be agreed with suppliers, wherever possible, for each ingredient used. The specification should include tolerances for desirable product attributes and acceptance quality limits for the levels of defects.

7.1.2   The implications of changes of raw material quality, within specification, on the effectiveness of the thermal process should be clearly understood.

7.2   WATER

7.2.1   **Potable Water**
See subsection 5.12.1.

7.2.2   **Potable Steam**
See subsection 5.14.

7.3   QUALITY ASSURANCE AND APPROVAL

7.3.1   Each ingredient should comply with its specification, and the control function should include knowledge of the suppliers' quality assurance programme, as well as a system of regular inspection of the ingredients and the removal or rejection of unsuitable material.

7.3.2   Deliveries of raw materials should be quarantined, where appropriate, until inspected, sampled, tested and released. Temporarily quarantined material should be located and/or marked in such a way as to avoid risk of its being accidentally used before release.

7.3.3   Ingredient materials should be delivered in packaging that maintains the integrity of both the ingredient and the environment and is itself not a source of contamination. The method of transport and delivery should be suitable for foodstuffs.

7.3.4   It may not necessarily be enough to assume that the description of a consignment of a raw material on the packages, on the corresponding invoice or on vendor assurance documentation is accurate Where the identity is not absolutely obvious beyond question, the identity of each consignment of raw material should be checked to verify that it is what it purports to be. The concern is that use of the wrong material or grade of ingredient may cause the food to be unsafe. Examples of raw materials of possible concern include "white powders" in poorly marked bags.

7.3.5   It should be noted that the traditional approach to ingredient quality control is being superseded by supplier (vendor) assurance programmes, in which the onus for checking product conformation with specification shifts to the supplier, who then guarantees to deliver product of suitable standard. Confidence in the supplier needs to be established, however, before the level of inspection on delivery may be reduced.

7.4   HANDLING AND STORAGE

7.4.1   All ingredient materials should be protected against contamination and should be handled hygienically and treated in such a way that deterioration is reduced to a minimum.

7.4.2   Facilities for bulk storage of ingredients should be of adequate capacity and constructed in a manner which will provide the protection indicated in subsection 7.4.1. All equipment used for the handling of ingredients should be of sound and hygienic construction (see subsection 8.1).

7.4.3   Each delivery or batch should be documented so that, if necessary, any batch of finished product can be correlated with the raw materials used in its manufacture and with the corresponding records. Deliveries should be stored and marked in such a way that their identities do not become lost.

7.4.4   Stocks of raw materials in store should be inspected regularly and sampled/tested, where appropriate, to ensure that they remain in acceptable condition.

7.4.5   In issuing raw material from store for production use, proper stock rotation should be employed.

8 PREPARATION PROCEDURES

8.1   Hygienic Design of Equipment and Hygienic Operation
8.2   Preparation of Ingredient Materials
8.3   Product Formulation and Reformation
8.4   Lot Traceability
8.5   Filling

8.1   HYGIENIC DESIGN OF EQUIPMENT AND HYGIENIC OPERATION

8.1.1   Equipment used in the preparation of heat preserved foods should be designed and constructed to be capable of hygienic operation so that the risk of microbiological or foreign material contamination of the food is minimized.

8.1.2   Equipment that comes into contact with food should be constructed of materials which are suitable for use with the foods and cleaning materials which may be used.

8.1.3   All equipment should be so sited that it is accessible for maintenance and cleaning purposes. Overhead fixtures, fittings and equipment should be maintained so that they do not present a contamination risk to the product from rust, flaking paint or dirt.

8.1.4   Equipment that remains in the production area, although not in use, should be cleaned and maintained in a hygienic condition and it should not cause overcrowding of the area.

8.1.5   Facilities should be provided, preferably in a specified area, for the cleaning and disinfection of containers, working implements and other small items of equipment. Such facilities should be provided with a suitable and adequate supply of water and adequate means of disposal of waste water.

8.1.6   Utensils or containers used for inedible or contaminated materials waste should be clearly marked and used only for their defined purpose. They should be sited away from open food and cleaned in an area specifically reserved for this purpose. Such containers should be removed at suitable intervals so that overloading and subsequent spillage is prevented.

8.1.7   Process equipment should be inspected regularly to ensure that all components are entire and maintained in a hygienic condition.

8.1.8   High standards of hygiene should be achieved and maintained, particularly in meat, fish and poultry preparation areas. Special attention should be given to the efficacy and frequency of the cleaning of the equipment and the manufacturing area to minimize contamination of meat, fish and poultry.

8.1.9   Adequate refrigeration facilities should be provided for the storage of chilled and frozen meat, fish and poultry. All such facilities should be kept clean and maintained in good repair. Temperature indicating or recording devices or both should be provided on all refrigeration storage units, and frequent checks should be made to ensure that the specified temperatures are being maintained.

8.1.10   Every effort should be made to keep the temperature for fresh meat, fish and poultry below 70°C up to the point where it is cooked or filled into containers. Prepared meat, fish and poultry not required for immediate use should be stored either chilled or frozen.

8.1.11   Where raw meat, fish and poultry are cooked prior to filling, the cooked product should either be filled hot or reduced quickly and without delay to a temperature below 7°C and held at this temperature until required for filling. The scheduled process should specify the temperature of the product at filling.

8.1.12   Cutting knives and similar instruments are a source of considerable contamination during preparation of meat, fish and poultry. Colour coding of knives should be used to designate use for either raw or cooked material preparation – but not both. Facilities should be provided for cleaning preparation equipment and surfaces with hot water at a temperature of 82°C (180°F) or above. Small items of equipment should be cleaned frequently during the course of the working day.

8.1.13   Equipment used for raw meat, fish and poultry preparation and that for cooked products prior to filling should be of stainless steel or other smooth, non-absorbent, corrosion-proof material. Cutting boards should be of non-absorbent materials and maintained with a smooth surface. They should be renewed at appropriate intervals.

8.1.14   Adequate precautions should be taken to prevent contamination of the prepared ingredients prior to filling. Covered receptacles should be used for transporting the product within the establishment.

8.1.15   Purpose designed, easily cleaned receptacles maintained in sound condition should be provided for waste materials and suitably located. They should be emptied frequently and subjected to a daily cleaning process.

8.2   PREPARATION OF INGREDIENT MATERIALS

8.2.1   All stages of preparation of ingredient materials for heat processing should be carried out under satisfactory conditions of hygiene and sanitation.

8.2.2   Adequate washing is an essential preliminary in the heat processing of vegetables and some fruits.

8.2.3   In addition to washing of ingredient materials, there may also be requirement for inspection and sorting in order to remove defective items or contaminants. These processes may be manual or mechanical in nature. If unexpected extraneous items are found, it is advisable to attempt to determine their origin in order to prevent recurrence.

8.2.4   In order to avoid microbiological cross-contamination, it is desirable that the preparation of meat, fish and poultry or other materials should be carried out in separate areas (see subsections 5.3.1 and 5.8.1).

8.3 PRODUCT FORMULATION AND REFORMULATION

8.3.1 Product formulation is a critical factor in achieving lethality during thermal processing and which must be specified in the scheduled process. If, for any reason, the formulation or method of product make-up is changed, full consideration must be given to deciding whether the thermal process requires to be amended. In case of uncertainty, the thermal process must be re-evaluated and, where necessary, re-specified.

8.3.2 Relative proportions of ingredients within each container should be controlled within known limits. Changes in the product composition or in the size or shape of an ingredient material may necessitate re-determination of the scheduled process.

8.3.3 If dehydrated ingredients are used in recipes, these may be wholly or partially reconstituted prior to filling. It is important that the scheduled process is established to take account of their condition at the time of sterilising. Similar logic applies to the use of frozen ingredients, where initial product temperature may be affected by the state of the raw material.

8.4 LOT TRACEABILITY

Product liability in the food industry requires that there be an adequate system of identification and traceability of lot. This is to ensure that a product recall system may be satisfactorily operated. It is necessary that finished product may be ultimately related, as closely as practically possible, to the raw materials, process conditions and containers from which it was made.

8.5 FILLING

8.5.1 Product quality and safety require that the preparation, filling, closure and commencement of final heat processing shall be carried out in an efficient manner, with no undue delay in the process which will allow growth of micro-organisms. In the event of breakdown or unforeseen delay, the effect of those delays on the product and on the efficacy of the scheduled process should be carefully considered, and action taken according to the circumstances to ensure the integrity and safety of the product.

8.5.2   All empty containers should be stored and handled in such a manner as to minimize damage and contamination.

8.5.3   Containers conveyed to the filler should be inverted and cleaned immediately prior to filling with a suitable air or water jet cleaning device. Water cleaning should not be employed for:

a. Containers used on aseptic filling lines, unless they are completely dried before sterilisation;
b. Heat sealed containers, owing to the risk of seal area contamination.

8.5.4   Containers should be filled in such a manner that the minimum of air is entrapped in the material being filled. Lack of control in the degree of air entrapment could have significant consequences for the effectiveness of sterilisation and for product quality.

8.5.5   If heat penetration is achieved by agitation of the product during the heating process, it is essential that the headspace limits specified in the scheduled process are maintained. Checks should be made regularly throughout each production period and the data recorded.

8.5.6   When the pack consists of solids in a liquid medium, the weights of solid and liquid components should be checked frequently and in-going weights adjusted if necessary.

8.5.7   Product filling temperatures should be maintained within specified limits. Gross variations in the filling temperature may result in failure to achieve the required degree of sterilisation for the scheduled process. The temperature at the filling machine, if part of the scheduled process, should be checked at intervals throughout each production period and the results recorded.

8.5.8   The initial temperature, i.e. the product temperature at the commencement of sterilisation, will be specified in the scheduled process. It is controlled effectively by control of the filling temperature and by subsequent control of the maximum hold period between filling and the start of sterilisation.

9 CONTAINERS AND CONTAINER CLOSURE

9.1   INTRODUCTION (CONTAINERS)

Containers for low acid foods include cylindrical and non-cylindrical tin-plate or aluminium cans, glass jars, semi-rigid plastic containers, semi-rigid foil-based containers and flexible pouches. Depending on the materials of construction, there may be upper limits to the retorting temperatures suitable for some containers.

9.1.1   Packaging materials should conform both to legislative requirements for materials in contact with food and to the quality standards within the specification agreed with the packaging manufacturer. An appropriate sample should be examined from each incoming consignment to confirm compliance with specification and to detect evidence of transport damage. The incoming sampling procedure should be agreed with the container manufacturer, and the extent of checking upon delivery will depend on the comprehensiveness of the vendor's quality management system.

9.1.2   Packaging materials should be handled and stored at all times with care to prevent damage likely to compromise the final integrity or suitability of the material. Empty containers and closures should be stored in clean and dry conditions. Stock should be issued on a strictly rotational basis. Records should be kept which will allow finished product to be reconciled with batches of packaging materials.

9.1.3   Storage conditions must ensure that taints are not imparted into packaging materials or containers during the storage period.

9.2   INTRODUCTION (CLOSING)

Control of closing operations (e.g. double seaming, capping, heat sealing) is essential to maintain product safety and quality. The objective is to obtain hermetic seals reproducibly with specific dimensions or characteristics within defined tolerances.

Closing and sealing machines must be in good condition, properly set and maintained to the manufacturer's specification, and closing and sealing operations should be under constant supervision.

9.3   CANS AND ENDS

9.3.1   Cans and ends represent critical elements in the safe production of ambient stable foods. Failure of container integrity at any point in the storage life may lead to microbial ingress, including pathogens, and product spoilage.

9.3.2   Cans and ends should therefore only be purchased from reputable suppliers able to provide competent technical support with regard to the use of their containers.

9.3.3   It is important that the can specification used is compatible with the product to be contained and the desired shelf-life. In the first instance the can supplier should advise on the suitability of the can specification for its desired purpose. During factory operation it is most important that the usage of cans is controlled to ensure that cans are used for the filling of products compatible with their specification.

9.3.4   Cans should be protected during storage in such a manner to prevent corrosion or mechanical damage, especially to the flange areas of the cans, and to prevent the ingress of foreign material or insects into the containers.

9.3.5   Cans conveyed to a filling machine should be inverted and cleaned immediately prior to filling with a suitable air or water jet cleaning device (see subsection 8.5.3).

9.3.6   The can maker should assume primary responsibility for specifying the correct tolerances for can seam dimensions. It is a prerequisite for successful operation that appropriate written documentation is supplied to the food manufacturer by the can maker (see subsection 9.6 on double seam control procedures).

9.3.7   Cans should never be used for purposes other than the filling of their intended product(s).

9.4   DOUBLE SEAM FORMATION

9.4.1   A double seam may be defined as a hermetic joint formed by interlocking the edges of both the end component and body of a can. It is commonly produced in two operations and involves the use of a pre-placed sealing compound on the end component.

9.4.2   The first operation achieves the initial interlock of end curl over can flange, with the second operation compressing the interlocked metal flat against the can body wall. Each can and end type uses a specific design of seaming roll groove profile which generates the optimum seam shape and dimension for the components being double seamed. Inadequate first operation cannot be corrected by subsequent second operation setting.

9.4.3   There are two basic designs of double seaming machines:

a. Stationary type, where the can and end are held stationary and the seaming rolls rotate around the components. Operating speeds of 10–200 cpm are achieved.
b. Rotating type, in which the double seam is formed when the can and end rotate whilst clamped between rotating components. With multiple seaming stations (from 3–18 heads), operating speeds of 100–2000 cpm can be achieved.

Double seaming machines may also be designed to produce specific headspace conditions and may operate under atmospheric, steam flow, cold vacuum or undercover gassing conditions.

9.5   DOUBLE SEAMING MACHINES – MAINTENANCE

9.5.1   Double seaming machine maintenance programmes should include the following:

a. Daily cleaning at the end of production and lubrication using food grade grease and oils.
b. Weekly checks of: seaming rolls and chucks for wear, damage and correct alignment; machine target timing and guide settings; end feed systems for cleanliness, damage and correct settings.
c. Monthly checks of: pin height setting; lifter spring loads.
d. Annual maintenance by a competent engineer familiar with the seamer in use.

9.5.2   Routine and planned maintenance and setting of double seaming machines will help to minimize the risk of problems during production. During maintenance checks all settings should be recorded before and after adjustment to highlight potential problems. The date and reason for fitting new chucks and rolls should also be recorded.

9.6   DOUBLE SEAM CONTROL PROCEDURES

9.6.1   Examination of double seams for visual defects should be undertaken throughout production, with especial attention after any machine jam or adjustment to ensure the absence of major deficiencies, e.g. cutover, knocked down flange, spurs.

9.6.2   Samples for detailed examination should be taken at regular and frequent intervals for each seaming head, but especially before production commences, after a significant stoppage or after adjustments are made to the seaming equipment. All pertinent observations should be recorded. The frequency of sampling (normally at least every four hours) will depend on the type of can, the closing machine and line speed, but this interval may be extended provided established data demonstrates double seams produced are consistently within the tolerances specified by the can manufacturer.

9.6.3   The evaluation of double seams should be carried out by competent personnel and should follow recognised procedures. When abnormalities are found, the corrective actions taken should be recorded. Both individual measurements and trends are useful in the assessment of double seam quality control and tooling wear. Instruments used for double seam evaluation should be suitably calibrated.

9.6.4   When measuring double seams on three-piece cylindrical cans, two points approximately 60 degrees either side of the side seam should be measured. In the case of drawn cans, two measurements should be made at opposite sides. For rectangular cans, a minimum of six separate measurements should be made, one at each corner and one at each long side.

9.6.5   Seam projectors provide a rapid means for obtaining relevant seam dimensions but only the tear-down procedure provides information on tightness grading (wrinkling).

9.6.6   The following dimensions are recorded (see Figure 1):

a. Prior to tearing down the double seam:
   – Countersink depth (e)
   – Double seam length (f)
   – Double seam thickness (h).

b. After tearing down the double seam or removing a section of the seam:
   – Body plate thickness (Tb) (Body wall thickness)

- End plate thickness (Te) (End component thickness)
- Body hook length (d)
- End hook length (g).

  The following are calculated or assessed (see Figure 1):
- Actual freespace
- Actual overlap (a)
- Per cent body hook butting
- Internal chuck wall impression
- Internal droop
- Tightness (wrinkles, pleats and other irregularities).

$$\text{Freespace} = h - (3Te + 2Tb)$$

$$\text{Actual overlap} = a = (g + d + 1.1Te) - f$$

$$\text{Per cent body hood butting} = \frac{b}{c} \times 100 = \frac{d - 1.1Tb}{f - 1.1\,(2Te + Tb)}$$

Evaluation of internal chuck impression, internal droop and tightness should be carried out with reference to the can manufacturer's double seam manual.

Overlap, body hook and end hook lengths may be measured directly from the optical projection of a double seam section. Overlap may also be calculated from the other measured parameters in the absence of a seam projector.

9.6.7 The container suppliers should provide specified acceptable values for the measured parameters of double seams for their containers. Typically acceptable figures provided by the Metal Packaging Manufacturers Association (MPMA) for three-piece tinplate cylindrical containers with 73 mm type B ends are:

Critical parameters:

Tightness rating		70% (minimum)
a	Actual overlap	1.02 mm (minimum)
b/c	Body hook butting = b/c x 100	70% (minimum)

Other parameters:

d	Body hook length	2.03 ± 0.13 mm
e	Countersink depth	3.18 ± 0.13 mm
f	Seam length	2.79 – 3.05 mm
g	End hook length	2.03 ± 0.13 mm

Note: Dimensions based on the average of two points of measurement 2 and 10 o'clock (side seam 12 o'clock) for three-piece and diametrically opposed for two-piece cans. Tightness rating – based on assessment of worst position on hook.

9.6.8   Non-cylindrical cans require special consideration, and the can manufacturer's specifications should be consulted and carefully followed to ensure that appropriate measurements and observations are made at the critical locations. In the case of rectangular cans, seam measurements are normally made at each corner and at the centre of each long side.

9.6.9   Certain plastic containers may be closed by double seaming. Whilst, in general, the construction of the double seam and its control and measurement are similar to that of double seams on all metal cans, there may be differences in specifications and critical locations. The container manufacturer's specifications should be consulted and carefully followed to ensure that appropriate measurements and observations are made at the critical locations. Tear-down procedure is generally inappropriate for plastic containers due to the flexibility of the plastic body hook. Seam projection is therefore necessary to obtain the required dimensional information of the double seam.

9.7   GLASS CONTAINERS – GLASS BREAKAGE PROCEDURES AND SPECIAL HANDLING CONSIDERATIONS

9.7.1   Glass Breakage Procedures and Special Handling – Repetition Considerations

Because of the hazards associated with glass contamination of foodstuffs, it is essential that any company packing food in glass takes suitable precautions to minimize the risk by line design, careful installation and operation, and the use of formal documented procedures to deal with any glass breakages which do occur.

a. Lines should be designed to control jar impacts, including when there is a line stoppage. Similarly, line stoppage should not result in jars falling over or falling off the line. Lines should, where practicable, be adequately screened such that if breakage does occur, the broken glass is contained and does not contaminate other parts of the factory. In particular, glass cleaning and filling equipment must be suitably screened, and it is good practice to fully enclose all conveyors between cleaning and closing. Additionally, conveyors for glass should not pass over areas where exposed food may be held.

b. Pallets of glass jars should be checked on delivery to the factory for the presence of broken glass, which, if present, should be a cause for rejection. Pallets with broken jars, if not returned to the supplier, should be held in a segregated area and sorted under suitable lighting and supervision.

c. Suitable lidded containers to be used only for the disposal of broken glass must be provided at intervals along the line.

d. All breakages must be cleared up immediately, preferably using industrial vacuum cleaners. Where appropriate, trained staff may use soft brushes and low pressure water.

e. Records of glass breakages should be kept to show where the breakage took place, time and length of stoppage, cause of breakage, management confirmation that the glass breakage procedure was followed, and that the line had positive approval to restart.

f. Specific glass breakage procedures should be documented for each particular part of the line listing all actions to be carried out, including the numbers of jars to be removed, cleaning procedure, test running, replacement of screens and logging of information.

g. Particular care and expertise are required for the cleaning of fillers, cappers and jar cleaning equipment to ensure that all particles of glass are removed. With fillers where glass fragments may have entered the filling heads, these must be flushed out and at least one rotation of the filler made and the filled containers discarded.

h. When glass breakage occurs between jar cleaning and closing, particles of glass may contaminate other open jars. Any jars which could have been contaminated must be removed from the line and discarded if already filled or put back through the jar cleaner if empty. The speed of the line and the time taken for it to stop are both important, e.g. if line speed is 300 jars per minute and is stopped 40 seconds after the breakage, at least 200 jars must be removed downstream of the breakage, together with those at risk in the breakage area.

9.7.2   With jars, the hermetic seal is normally created between the lining compound within the closure and the sealing surface across the neck of the container itself. Closures for glass containers for food may be divided into two broad categories, namely venting closures and non-venting closures.

9.7.3   A venting closure may be defined as a closure which is simply crimped onto the jar and which then permits, by controlled venting, the escape of air entrapped in the jar during subsequent heat treatment. With the onset of the cooling phase, a vacuum is created in the jar, this vacuum pulling the closure down and thereby resealing it on the neck of the jar.

9.7.4   By contrast, a non-venting closure is applied to the jar under vacuum, thereby giving an immediate hermetic seal, and the closure is maintained in this fully sealed condition throughout subsequent heat treatments.

9.8   CAP AND GLASS SELECTION

9.8.1   It is important to select the appropriate type of both cap and container for high temperature processing. Knowledge of the products to be packed and the process treatments to be given is necessary so that appropriate internal coatings and the appropriate lining compound may be applied to the closure.

9.8.2   With regard to the selection of the container itself, it is important to use the glass neck ring finish appropriate for the chosen closure, and to ensure that any glass surface treatment used is discussed with closure suppliers, as it is known that excessive levels of certain treatments do have an adverse effect on removal torques and capping performance.

9.8.3   It is most important when selecting a design for the glass container to be used that the skirt of the closure does not project beyond the shoulder or body of the container. This is necessary in order to prevent damage to the skirt of the closure during handling and processing. Any damage in this area could lead to the disturbance of the cap and possible reduction in closure integrity and vacuum retention.

9.9 FILLING AND CAPPING OF GLASS CONTAINERS

With regard to the packing line and processing operation itself, there are six areas which are most important in ensuring that the containers are hermetically sealed and capable of being successfully processed and stored, namely:

a. Filling conditions (see subsection 9.9.1);
b. Capping conditions (see subsection 9.9.2);
c. Filled container handling prior to processing (see subsection 9.10);
d. Processing conditions and equipment being used (see subsection 9.11);
e. Filled container handling after processing (see subsection 9.13);
f. Subsequent warehouse storage of the filled stock (see subsection 9.16).

The following considerations apply principally to the non-venting type of closure but are important for the venting type of closure also – particularly subsections 9.11.3–9.11.7.

9.9.1 Filling Conditions for Glass Containers

When filling glass containers with food products, it is important to achieve a consistent headspace and to avoid soiling of the sealing areas. This is usually achieved by attention to:

a. Clean filling conditions maintained in order to prevent contamination of the sealing surface of the container.
b. Product filling temperature maintained within a consistent range. The product temperature used has a significant influence on the build-up of pressure within the container during high temperature processing.
c. Consistent filling level. The headspace volume within the filled containers must be adequate to prevent excessive pressure build-up during processing.
d. A flat product surface without projections above the container rim and with the minimum of entrapped air present.
e. Minimization of external product contamination of the walls of the jar. Such contamination could cause loss of grip between the side belts of the capping machine

and the containers, thereby reducing security of cap application.

f. Glass containers and caps should be designed to avoid cap-to-cap contact during conveying operations.

Regular checks should be carried out to ensure that the above conditions are maintained consistently and within acceptable limits.

9.9.2　**Capping Conditions**
It is most important that the setting up and operating procedures of capping equipment suppliers are adhered to when closing glass containers filled with food products.

9.9.3　It is necessary to control the performance of the capper, and the most important factors in assessing this performance with non-venting closures are:

a. A cold water vacuum check, which entails filling containers with cold water to a nominal 10–12 mm headspace, closing the jars and measuring the internal vacuum – a minimum of 500 mm Hg should be achieved.

b. A vacuum check on product filled containers. Normally vacua in the range 200–400 mm Hg are achieved in such containers, depending on product temperature and headspace being achieved, and consistency in such vacuum levels is an important factor in the scheduled process.

c. A visual check of the sealed containers themselves to establish that the closure is sealing correctly on the glass finish, and with a satisfactory impression being achieved in the lining compound throughout the 360 degrees of the cap.

d. A security check, which is a measurement of the tightness of application used with lug-type closures.

9.9.4　These checks are carried out on jars taken from the exit of the capping unit. The pressure and flow of steam to the capping machine is important as this both evacuates the air from the headspace of the jars and softens the lining compound in the caps just prior to application. However, when handling certain food products, e.g. some mayonnaise and other cold filled products, where the possibility of steam condensation within the headspace of the closed containers could initiate yeast or mould growth, either superheated steam (which is dry) or a cold closing technique using a mechanical vacuum capper may be utilised.

9.9.5   Reject containers after capping should be examined so that corrective action to remedy the causes of the defect may be taken.

9.10   FILLED GLASS CONTAINERS: HANDLING PRIOR TO PROCESSING

Once the containers have been filled and closed satisfactorily, they should be handled with care to avoid damage to caps which could result in subsequent leakage and spoiled packs. The following points are of particular importance during the handling of the containers from the end of the capper to the processing stage:

a. Side guide-rails on runways, conveyors, accumulators etc should be set at such a height that they support and contact the body and shoulder of the container; they should not make contact with the skirt of the cap.

b. The speed of conveyors should be synchronised to maintain container spacing in a order to eliminate container-to-container contact and thus prevent glass breakage or cap-tap contact, which could affect closure integrity.

c. When horizontal or vertical retorts are being used for processing, care must be taken when loading the retort crates in order to minimize abuse. When the containers are being loaded in layers into the crates by a manual sweep, speed of transfer of the containers should be reduced to the minimum necessary. If automatic crate loading is being carried out, the system must be adjusted to avoid cap and container damage when the bars are descending or advancing.

9.11   GLASS CONTAINERS:
PROCESSING CONDITIONS AND EQUIPMENT

9.11.1   Care must be taken to ensure that no unacceptable stresses are placed on the closure during heat processing as these could reduce the security of the seal by displacement of the closure or disturbance of the lining compound when it is soft and susceptible.

9.11.2   In order to maintain closure integrity throughout processing and initial cooling, it is necessary to ensure that an external pressure is applied which is greater than the internal container pressure developed. This overpressure is achieved by the application of either air or steam pressure within the retort, the actual method depending upon the type of retort system in use.

9.11.3   The internal container pressure during processing is governed by a combination of headspace volume, closing conditions, pack dimensions and the processing temperature used. The maximum internal pressure developed can be established by measurement, observation or calculation.

9.11.4   Calculation of internal pressure requires knowledge of:

a. The internal vacuum in product-filled containers;
b. The product filling temperature;
c. The percentage headspace of the filled containers;
d. The process temperature used.

9.11.5   During filling, processing and cooling operations, especial care must be taken to:

a. Avoid thermal shock from large temperature differentials between container contents and heating/cooling media;
b. Maintain a lower pressure in the container than the external environment at all times so as to retain the closure in position.

9.11.6   An overpressure should be applied as soon as the processing chamber is closed and the processing medium is introduced, and should be maintained throughout processing and initial cooling. The nominal overpressure should be maintained during the initial period of cooling, after which, based on experience, it can be reduced as necessary to complete the cooling cycle. Fluctuation around the nominal overpressure should not exceed 0.2 kg/cm^2 (3 psi) at any time.

9.11.7   The closure manufacturer's recommendations should be closely followed with regard to maximum overpressure and processing temperature as different values may be appropriate to different closures.

9.11.8   Maximum pressure during processing normally occurs during initiation of the cooling cycle and especial care must be taken at this stage. Excessive overpressure may lead to cut-through of the closure lining compound or, where relevant, could interfere with button operation.

9.12   SPECIAL CONDITIONS APPLYING TO GLASS CONTAINER RETORT (NON-CONTINUOUS) PROCESSING OPERATIONS

9.12.1   When using horizontal or vertical retorts as opposed to continuous cookers, the individual containers need to be loaded into retort crates for processing and care is required when this operation is performed. When stacking containers into crates to be processed in vertical retorts, it is also important to ensure that they are not being stacked to a level higher than the rim or top supports of the crate, otherwise top loading damage may occur due to the pressure exerted upon the containers by the crate(s) above. The design of the retort crate handles must also be checked to ensure that there is no ancillary damage arising from the mechanical handling of the filled crates.

9.12.2   Perforated plastics or rubber layer pads should be used to segregate the layers of containers, and these should be kept clean and in good condition so that they do not scuff the caps on the jars beneath.

9.12.3   During retort processing operations, it is necessary to avoid subjecting the glass containers to physical or thermal shock. To this end, in an immersion water system the containers should be sufficiently covered with water. There should be a cushion of air in the headspace of the vessel to prevent sudden changes of pressure, and water circulation should ensure even temperature distribution. The temperature differential between the containers and the process medium should not be too great at the outset of processing and the introduction of cooling water at the end of sterilisation needs to be carefully controlled.

9.13   FILLED GLASS CONTAINER HANDLING AFTER PROCESSING

See also sections 17.7 and 17.10.

9.13.1   Processed containers should be handled with care to avoid abuse which could result in subsequent leakage and spoiled packs. It is necessary to carry out routine checks on the processed containers in order to confirm that satisfactory processing conditions have been employed and that no disturbance of the closure has arisen during processing. The following points are of particular importance during the handling of the containers subsequent to processing and during warehouse storage of the palletised stock.

9.13.2   Side guide-rails on runways, conveyors, accumulators etc should be set at such a height that they support and contact the body and shoulder of the container; they should not make contact with the skirt of the cap.

9.13.3   The speed of conveyors should be synchronised to maintain container spacing to eliminate container-to-container contact and thus prevent glass breakage or cap-to-cap contact.

9.13.4   When horizontal or vertical retorts have been used for processing, care must be taken when unloading the retort crates in order to minimize abuse. Wet containers must not be manually handled because of the risk of microbiological contamination. When automatic crate unloading is being carried out, the pusher bars must be adjusted to avoid cap damage. The top layer of containers in the retort crates should be carefully examined to establish whether mechanical damage has occurred during handling of the crates into and out of the retorts.

9.13.5   Caps/containers must be dried as quickly as possible to minimize the risk of microbiological spoilage and external rusting of the caps. If the temperature of the product within the container is insufficient to allow them to dry themselves rapidly, arrangements must be made on the post-processing line to remove excess water remaining on the caps (particularly around the skirt (and in this respect jars are different to cans)) by the installation of an efficient drier or by the positioning of high velocity air jets at strategic points on the line. The design and disposition of the air jets in drying units is very critical in order to achieve maximum effectiveness.

9.13.6   It is recommended that low vacuum detectors be installed in the line after processing so that nil vacuum or low vacuum containers are eliminated. A regular examination of all rejected containers should be carried out in order to establish the reasons for failure, so that corrective action may be taken to overcome the problem. Detectors should be checked at intervals to ensure that they operate to required levels.

9.14   GLASS CONTAINERS: POST-PROCESS EVALUATION OF CLOSURE SEAL INTEGRITY

9.14.1   Reference has already been made to the range of examinations which should be carried out on non-venting glass containers immediately after capping (see subsection 9.9.3); it is also necessary to carry out a series of checks on the container/closures after heat processing in order to confirm seal integrity.

9.14.2   The range of evaluations which should be carried out on containers cooled to 20–25°C are listed below:

   a. A visual check to establish that the closure has been correctly applied to the glass container, and that there is a satisfactory impression in the lining compound of the closure. This lining compound impression must be consistent across the sealing surface of the container, and there must be no sign of compound thinning or cut-through. There should be no evidence of cracks or defects in the glass, nor of impact damage on the container/closure. Required information regarding batch number or shelf-life declarations should be legible.
   b. Vacuum levels in containers should be consistent, and should agree with the calculated levels based on experience of filling/closing/processing conditions.
   c. The security check should be performed on lug-type closures, and is designed to confirm that the caps have been applied correctly. Having established a satisfactory security specification under the packing line conditions being used, regular checks will indicate whether the closures are being over, under or correctly applied.
   d. Opening torque readings should be consistent and should be within the range of readings established with the closure manufacturer. Opening torques will normally increase on storage, and it is common practice

to carry out a further examination after 2-3 weeks' storage.

e. On those closures which incorporate a button feature, the button must perform satisfactorily on opening (vacuum release).

f. It is recommended that a low vacuum detector should be installed within the packing line. The performance of the low vacuum detector should be checked to confirm consistent rejection of nil vacuum/low vacuum containers. Rejected containers should be examined to establish the reason for failure, and a record of the reject rate should be maintained so that any trend towards an increasing failure rate may be highlighted and corrected.

9.15 TAMPER EVIDENT/LOW VACUUM CAPS

Vacuum button caps may be used to provide immediate visual indication of loss of vacuum within containers. The loss of vacuum may be due to faulty sealing, irregularities during processing or to mechanical tampering at some stage after processing.

9.16 WAREHOUSE STORAGE OF FILLED STOCK OF GLASS PACKAGED FOODS

9.16.1 The final outer packing operation and subsequent warehouse storage of filled stock is also important in maintaining the processed containers in optimum condition. They may be palletised for storage in one of two ways, namely in their final shrinkwraps or cartons, or alternatively as loose containers.

9.16.2 Care should be taken to minimize the disturbance of caps during initial storage in order to allow the sealing compound to stabilise. This can be achieved by adopting correct handling and storage practices, including those described below:

a. Where containers are being stored loose on pallets for a period, fluted fibreboard layer pads should be used between the layers of jars. Prior to the final labelling/shrinkwrapping/cartoning operation, containers should be subjected to a dud detector or tap-test examination to eliminate slow leaking packs caused by, for example, hairline cracks across the container sealing surface.

b. Where containers are packed into shrinkwraps rather than cartons for distribution, it is recommended that microflute board is used for the shrinkwrapped trays.

c. The pallets used must be in good condition and palletised loads must be lowered gently into position to minimize uneven distribution on the pallets beneath. It is also recommended that divider pads be placed between each stacked pallet to reduce still further the uneven distribution caused by single-faced pallets.

d. The stacking height of the pallets is significant with regard to possible top-loading abuse of the caps. As long as the above recommendations are followed, there would appear to be no significant increase in top loading abuse on pallets stacked three high in warehouses for long term storage, whereas four high stacking of pallets must be carefully controlled. Stacking of palletised containers more than four pallets high is not recommended.

9.17 SEMI-RIGID PLASTICS CONTAINERS

9.17.1 Semi-rigid plastics containers for heat processed foods are manufactured from combinations of thermoplastics, sometimes with other materials added to give additional properties. Materials are selected to withstand the thermal processes to which the foods they contain are to be subjected, including reheating by the consumer. Their selection is based on the properties required of the container, such as rigidity, flexibility, barrier properties, and the compatibility between closure, container and product. Advice should be sought from the packaging supplier in relation to suitability of containers to their intended purpose.

9.17.2 Plastics packaging materials should be delivered clean, and subsequently protected from contamination during storage. They should not be washed with water prior to use (unless effective drying is possible) because the presence of water in the sealing area may reduce seal reliability. Containers should be cleaned by blowing or vacuuming and inversion prior to filling.

9.17.3 Packaging materials should be stored so that they are protected from damage leading to cuts, cracks or distortions. Storage temperatures prior to use should not prejudice the formation of the heat seal nor alter the physical properties of the packaging materials. For example, cold storage may make some types of plastic brittle and fragile. Storage periods should not be so long that dimensional changes occur.

9.17.4   Rough handling of plastics containers during transport operations must at all times be avoided as the integrity of the hermetic seal, or of the body of the container itself, could be ultimately affected. Containers should be generally transported on their bases, and impacts with other containers or with guide-rails of conveyors kept to a minimum. Where containers are conveyed upside down, the container should be designed to ensure that the actual seal area IS NOT in contact with the conveyor.

9.17.5   Two main methods are used to provide hermetic closures to semi-rigid plastics containers. These are:

a. Heat seals. Heat is applied, or generated, to bring the sealing surfaces to a molten or semi-molten state, causing them to fuse together with the application of pressure. Some heat seals are designed to be peelable for easy opening.
b. Double seams. The double seam is formed by interlocking the curl of the metal end to the flange of the plastic body in a process similar to that used for conventional metal cans. The container supplier should specify the seam dimensions necessary to provide an hermetic seal.

9.17.6   Plastics containers are different from those made from metal or glass in that they soften appreciably upon the application of heat. Any pressure development within a container during sterilisation could therefore cause considerable container deformation. This may be of a permanent nature and lead to unsaleable containers, but equally importantly, the heat transfer characteristics inside the container could be so altered that understerilisation results. During thermal processing the designed processing temperature of the containers must not be exceeded and adequate pressure control on the vessel is vital. The packaging supplier should be consulted to advise on the overpressure required in order to minimize container deformation during processing.

9.18   HEAT SEALING OF SEMI-RIGID PLASTICS CONTAINERS

9.18.1   Filling plays a critical role in ensuring that hermetic seals can be formed and that pack integrity is maintained during the sterilising operation. Alignment and performance of the filling heads should be checked to ensure that the seal area is not damaged, distorted or soiled during filling. The quantity filled, including the product components affecting the rate of heat penetration, must be accurately controlled. This is especially important if the pack has a gas-filled headspace requiring the application of a compensating overpressure to avoid pack distortion and bursting during the sterilisation cycle.

9.18.2   The dimensions of the lid and base and the composition and thickness of the sealing layers are critical to sealing. These critical features should be tightly controlled by the packaging supplier and checked against appropriate acceptance criteria on delivery to the processor. Checking should include confirmation of proper components and the thickness of each component. The most critical packaging material dimensions are those affecting the strength, evenness and integrity of the seal. Checks should monitor especially the thickness, and variation in thickness, of the materials presented to the sealing head and any dimensions of the pack which may lead to seal misalignment.

9.18.3   The heat seal is formed by melting together two or more plastic layers in a defined area under precisely controlled conditions of time, temperature and pressure. Control systems on the sealing machine must reliably control critical aspects of seal formation. Sealing machines should ideally be fitted with alarms or stops which operate when the sealer is operating outside its control limits.

9.18.4   The critical features of the sealing operation should be identified before production starts. Target values and limits should be set for each part of the sealing operation, including container delivery and positioning, pre-sealing, any gas or steam flushing or vacuum cycles, the main sealing stage and any cutting operations. Sealing data should be recorded where appropriate.

9.18.5　Some sealing systems are able to seal across soiled seal areas. That they can reliably do this must be established experimentally for the expected combinations of sealing head (especially its profile), sealing conditions, composition of the fill and frequency and quantity of soil.

9.18.6　Measurement of the resulting pack and seal dimensions and strength can be used to check sealer performance. Different materials and sealing systems will produce seals and packs with different critical dimensions and attributes. Suitable specifications should be derived from discussions with equipment and packaging material suppliers. Seal width is generally an important feature – a normal target width is about 3 mm, with a minimum width of 1 mm. A list of some typical packaging tests is given below.

Typical tests for seal integrity:

a. On-line visual examination of packs and seals will only identify gross faults. More careful examination off-line can identify sealing faults such as pleats, wrinkles, bubbles, delamination, off-centre or uneven seals.
b. Peel test. The lid is peeled away from the base under controlled conditions. The force required to initiate and continue peeling can be noted (seal tensile strength), and the appearance of the peeled area on both surfaces should be noted to establish that an even, continuous seal has been formed.
c. Gas leakage. Packs are pressurised off-line and the loss of pressure over a period of time measured. This test is only effective if any holes in the seal or pack are not blocked by product.
d. Helium detection. A small amount of helium gas is added to the headspace gas. The container is subsequently subjected to vacuum and any helium escaping is detected using a mass spectrometer.
e. Dye penetration. The product side of seals is exposed, cleaned, dried and immersed in a dye solution. After a period of time, the outside of the seal, and then the seal itself, are examined to see whether any dye has passed from the product side to the outside, indicating that the seal is not hermetic.
f. Conductivity testing. The interior of the pack is filled with an electrolyte and then placed in a bath of electrolyte, so that the two electrolyte solutions are

separated by the packaging. The conductivity between the inside and outside of the pack is measured; high conductivity indicates that there is a channel across the seal or a hole in the pack.

g. Static biotesting is an off-line technique which examines the ability of the pack to exclude motile micro-organisms during immersion in a suspension of them. It is a severe test and is only suitable for use during the commissioning of equipment or for determining whether sealing faults are critical (i.e. lead to loss of pack integrity) or of less importance.

Tests for seal strength:

a. Burst test. The pack is pressurised to a designated pressure specified for the materials or the application (e.g. 0.8 bar) for a period of time (e.g. 30 seconds) to check the strength of the pack and seals. This test will not reliably detect whether the pack is hermetic or not, it is only designed to test seal strength. The type of pack failure may vary and should be noted.

b. Peel test (see b above).

c. Seal penetration. The overall reduction in thickness of material in the seal area (i.e. material which has been between the sealing jaws) is measured to determine the combined effects of sealing head pressure, jaw temperature and dwell time. This test can only be used reliably if the base and film thicknesses have been measured prior to sealing. When measurements are taken around the circumference of the seal, it provides a good indication of the system's overall performance.

d. Drop test. The outside of dry, sterilised packs is coated with bacterial spores, especially in the seal area, re-dried if necessary and then packaged, handled and palletised as normal. They are subjected to a standardised/representative set of mechanical shocks to simulate transport and distribution. Afterwards they are incubated to allow time for any of the spores which have reached the contents from the outside to grow and produce recognisable changes. Packs are then examined for these changes to indicate loss of integrity.

9.18.7 The frequency of container integrity testing, using some of the techniques described above, should be sufficient to ensure that consistently strong, hermetically sealed packs are produced. During routine production, the critical features of at least one pack per head or sealing lane should be examined by measurement and tear-down at every start-up and subsequently with a minimum frequency of once per 30 minutes. Visual inspection of packs should be done more frequently, e.g. every 15 minutes. Additionally, when there is a change-over in packaging material or where there is a join or splice in a film or web, a sample should be examined as at start-up. Similarly, when running adjustments are made to the sealing machine or the machine speed is altered, samples should be examined. Where there is any doubt over the integrity of a seal, all production since the last inspection should be isolated and representative seals examined by dye penetration to establish the importance of the fault.

9.18.8   After filling and sealing, the packs should be handled and conveyed so that stress and damage (for example by puncturing) are avoided. Holding periods and conditions should not allow significant growth of the microbial population of the unprocessed fill. Unpredictable changes in factors affecting heat penetration (such as viscosity) and a large range of initial temperatures in the load should be avoided, as these prejudice effective control of headspace volume during heating and cooling and weakened packs could result. Flexible and semi-rigid packs with a gas headspace must only be sterilised in retorts with effective overpressure control systems.

9.18.9   Racking systems or crates used to hold the packs in the retort should protect them from abrasion, puncturing and crushing, and should also ensure predictable heat penetration.

9.18.10   After sterilisation and cooling, packs should be taken out of the retort at a sufficiently high temperature to promote drying, but the temperature must be sufficiently low to prevent pack distortion by residual internal pressure. The cooling rate and the overpressure in the retort must be controlled to prevent pack seals being weakened by excessive negative or positive pressure in the headspace.

9.18.11   Packs should not be handled manually when they are wet. With many types of pack there will be surface water, which drains off rapidly or can be blown off, but water trapped in the seal area, which is held by capillarity, can only be lost by evaporation. Pack contamination through seal faults by this water in the seal area is a critical hazard and, for this reason, the packs must be handled under hygienic conditions until completely dry. Overwrapping of packs should not be done if it is likely that microbial contamination of the residual water in the seal area will result.

9.18.12   If the packaging system used produces packs that can be damaged by the stresses and impacts experienced during normal handling, they should be enclosed in a secondary, protective overwrap. Alternatively, multiple unit overwrapping and palletisation systems may be designed to ensure that the packs are able to resist such damage without the necessity of individual unit overwrapping.

9.18.13   Filled stock in warehouses should be held under conditions ensuring the retention of pack integrity and preventing damage to or weakening of packs; for example, by stacking pallets too high. If the strength of packs is reduced by temperature changes, then storage and transport should be under controlled conditions.

10 THE THERMAL PROCESS

10.1   The Principles of The Thermal Process
10.2   The Scheduled Process
10.3   Temperature Distribution Tests
10.4   Heat Penetration Tests
10.5   Evaluation of Heat Penetration Data

10.1   THE PRINCIPLE OF THE THERMAL PROCESS

10.1.1   Thermal processes applied to the preservation of foods in cans and other hermetically sealed containers are intended to destroy all organisms capable of growth during subsequent storage of the food. The thermal process required to achieve this aim will depend on the composition of the food, the numbers and types of bacterial spores present, pH value, water activity, amounts and types of any curing salts or other microbial inhibitors present, and the expected storage conditions of the product. However, it must be remembered that the prime microbiological requirement for the safety of any heat preserved food is that it should be free of pathogens capable of growth in the food.

10.1.2   When considering the thermal process required for the safety and stability of canned foods, there are two main categories: LOW ACID products with pH values of above 4.5, and HIGH ACID products with pH values of 4.5 or below. Only LOW ACID products are considered in this chapter.

10.1.3   Low acid heat preserved foods which are able to support the growth of food poisoning micro-organisms must be given a heat treatment such that these organisms are killed. The pathogenic micro-organism with the most heat resistant spores is *Clostridium botulinum*, the organism which causes botulism.

10.1.4   There is no absolute guarantee that all micro-organisms will be destroyed by a heat process and, theoretically, there can be no absolute protection against *Cl. botulinum* in low acid heat processed foods. However, it is possible to work to a very low probability of microbial survival. For *Cl. botulinum.* a probability of survival of not more than 1 in 10^{12} containers is regarded as acceptable. This is usually interpreted as requiring a minimum thermal process value $F_0 = 3$ min for every container in the production batch. In practice, because of variations experienced in manufacturing and the need to control the heat resistant spoilage organisms, it is normal to aim for an F_0 value in excess of 3.

10.1.5   The minimum thermal process value may be derived using the following equation, assuming instantaneous heating and cooling:

$$F = D (\log N - \log S)$$

where   F = The equivalent time in min at a reference temperature
D = The decimal reduction time, i.e. time to reduce the bacterial spore population by 90% at the reference temperature and in a specific substrate
N = Initial number of spores
S = Number of surviving spores.

The most heat resistant spores of *Cl. botulinum* have a $D_{121.1}$ ($D_{250°F}$) of 0.21 min.

If 100 spores are present initially, the reference temperature is 121.1°C and the required probability of survival is not more than 1 in 10^{12}, then:

$$F = 0.21 (\log 100 - \log 10^{12})$$
$$= 0.21 (2 + 12)$$
$$= 2.94 \text{ min}$$

This time is usually rounded up to an F value of 3 min at 121.1°C.

Where the F value refers to a reference temperature of 121.1°C and a temperature coefficient of destruction (z value) of 10°C (18°F), the value notation is usually written as "F_o."

10.1.6   Microbiologically stable packs may contain small numbers
          of surviving micro organisms, usually in the spore form.
          However, such organisms are of neither public health nor
          spoilage concern as they are unable to grow. In recognition
          of this fact, such foods are frequently called
          "commercially sterile."

  10.2   THE SCHEDULED PROCESS

10.2.1   A scheduled process must be established for each low
          acid product and for each container size in which it is
          produced and for each type of steriliser used. The
          thermal process received by cans of low acid foods must
          be adequate to ensure the safety of those packs against
          the survival and growth of the spores of the organism
          Cl. botulinum. The process should also be adequate to
          prevent spoilage by other heat resistant, non-pathogenic
          organisms under the conditions of subsequent storage.
          Ingredient materials are sources of bacterial spores and a
          knowledge of the spore load is important in the
          establishment of the process. The scheduled process
          should be established by suitably trained personnel.

10.2.2   While heat penetration test information is an essential
          prerequisite, a scheduled process will be defined by a
          number of critical factors peculiar to the product and the
          sterilisation system employed, and not just the time and
          temperature of the sterilisation procedure. A typical
          scheduled process should include, but not necessarily be
          limited to, the following headings or critical factors:

          a. Product name, code, type, and formulation reference;
          b. Product characteristics, including particle size and pH;
          c. Container size and type;
          d. Container fill weight;
          e. Headspace in the container;
          f. Orientation of the container, with/without nesting;
          g. Minimum initial product temperature;
          h. Pre-process hold time (minimum/maximum);
          i. Type of heat processing system;
          j. Sterilisation temperature;
          k. Sterilisation time;
          l. Specific come-up or pre-cooling conditions, where
             applicable;
          m. Container rotation, where applicable;
          n. Overpressure, where applicable;
          o. Cooling method and its time and temperature;

p. Conveyor speed, can agitation, where applicable;

q. Venting procedure, where applicable;

r. The date on which the schedule was established.

Changes in factors which form part of the scheduled process must be verified by authorised expert personnel before implementation. This may involve further heat penetration studies.

10.2.3 Heat penetration data for a product showing a simple (pure conduction) heating curve may be converted by a recognised mathematical method in order to calculate the scheduled process for changes such as a different can size, initial temperature or sterilisation temperature. However, such conversions must be verified by experiment at the earliest opportunity, before commercial production and before reliance is placed upon them.

10.2.4 It is often considered necessary with some types of foodstuffs to conduct incubation tests on both experimental and initial commercial production runs. The statistical and microbiological principles of the tests should be clearly understood. Incubation tests should NEVER be used as the sole or main criterion for assessing the safety of any heat process or the bacteriological status of any production batch of canned foods. (See section 18.)

10.2.5 Records giving full details of how and by whom the scheduled process was established and confirmed should be retained permanently on file. Any changes to the scheduled process should be kept on record with the original data. Records should be accessible to those concerned with the management of thermal processes.

10.3 TEMPERATURE DISTRIBUTION TESTS

10.3.1 Before a retort or sterilising system is brought into service and any heat penetration work undertaken, temperature distribution tests should be carried out in order to determine the slowest heating point in the sterilising vessel, i.e. that point that is slowest to reach the scheduled processing temperature, and also to confirm that the range of temperatures experienced throughout the retort is within prescribed limits. These tests should be repeated every three years (or more frequently if engineering changes have been made to services or controls) and results kept on file.

10.3.2   The temperature distribution tests are also used to verify the adequacy of the venting operation for sterilising vessels heated solely by saturated steam. In these systems, it is necessary to remove all air completely in order to eliminate the possibility of localised lowering of temperature. The importance of correct venting cannot be overstressed.

10.3.3   With continuous sterilisers, documentary proof must be available from the manufacturer or site commissioning team that adequate venting can be achieved and that temperatures are within ±0.5°C of the scheduled processing temperature before heat penetration tests are undertaken.

10.3.4   In steady-state operation, the temperature spread across the sterilising vessel should ideally be 1°C or less. However, when this degree of control is not achievable due to design or characteristics of the equipment, any deviation from the limit must be allowed for in the scheduled process.

10.4   HEAT PENETRATION TESTS

10.4.1   Conventional heat penetration tests monitor the time/temperature history of the product at the slowest heating point in the container. (It is important to note that this may not be at the geometric centre of the pack.) The results are converted into arbitrary units of lethal heat, which are summed over the whole of the process, to give the total lethal effect, which is usually expressed as the Fo value. The tests should be carried out in the slowest heating area of the sterilising vessel.

10.4.2   Where heat penetration tests are made using laboratory retorts or simulators, results should be verified in the production retort under commercial operating conditions.

10.4.3   In the case of laboratory simulators of continuous sterilisers, it is important that the temperature/time and pressure/time regimes closely mimic those in the production machines. The relationship between the performance of the simulator and production machines must be understood and used in assessing the thermal process.

10.4.4   The heat penetration tests must be carried out under the most adverse conditions that are likely to be encountered in production, e.g. at the lowest initial product temperature. A single heat penetration test will not be sufficient to cover the variations which can occur during the production process. (See also CFDRA Technical Manual No. 3 for further discussion.)

10.4.5   Given the factors affecting the thermal process, it is clear that the lowest thermal process value (Fo) is not easily predicted or assessed. Therefore, sufficient determinations should be made to establish the variability of the thermal process and to find the lowest thermal process value likely to be encountered under production conditions.

It is recommended as a minimum that three separate heat penetration tests, each of at least three replicates, be carried out to allow a scheduled process to be set. More than this will be required if there is an unexpectedly large spread of results.

10.4.6   The temperature measuring system used to monitor the rate of heat penetration should:

a. Have a total accuracy within $\pm0.5°C$ ($1°F$);
b. Not significantly affect the manner in which heat is transferred within the container;
c. Be regularly checked against a certified standard to ensure its accuracy.

10.5   EVALUATION OF HEAT PENETRATION DATA

10.5.1   When heat penetration data from the slowest heating point of the container has been obtained, the data must be evaluated to form part of the scheduled process. Such an evaluation must only be carried out by persons with suitable qualifications and experience.

10.5.2   Although, in principle, the lethality of a thermal process may be determined by microbiological methods using spores of test organisms, these methods yield data of limited application, and consequently are used for specific purposes where heat penetration tests are not practicable. In these cases, spores of non-*philus* pathogenic *Bacillus* species e.g. *B. stearothermophilus*, contained, for example, in capillary tubes or bulbs, may be positioned within the container, using various means, to assess the lethality of the process. The lethality is calculated from the numbers of survivors, and the heat resistance of the spores.

11 MANAGEMENT OF THERMAL PROCESSING

11.1   Whilst the basic principles of thermal processing are relatively simple, almost every raw material and manufacturing step has an effect. All this must be considered during the establishment of the scheduled heat process and allowances made for normal variability. It therefore becomes essential that abnormal variability should be eliminated (or at the very least recognised, and the technical personnel responsible for heat process control informed).

11.2   Changes in primary packaging and raw materials, even a change in raw material supplier, can have a dramatic effect on heat transfer, the types of micro-organisms present, and consequently on lethality achieved. Small changes in manufacturing procedures or engineering modifications, including those made during maintenance, can also have very significant effects on these factors and the application of the scheduled heat process. (See subsection 10.2.2.).

11.3   The practicalities of thermal process control are therefore complex and require firm, ongoing management control across the whole of the raw material supply, manufacturing and, in certain instances, distribution operations.

11.4   A designated and senior person with suitable training and experience should be the only person to assign or authorise changes to the scheduled heat process. For the purpose of these guidelines, such a person shall be referred to as the Thermal Process Manager. A management structure should be set up and suitable training given such that any changes in materials, procedures or structures are notified to the Thermal Process Manager. (See subsection 10.2.2.).

11.5   As part of his function, the Thermal Process Manager should ensure that audits are carried out at least annually on all factors which could affect the validity or application of the scheduled process. These audits may not necessarily involve measurement of sterilisation values, temperature distribution in the sterilisers or bulk incubation tests, but should be designed to check that all changes to critical factors which have occurred have been notified to the Thermal Process Manager (see subsection 10.2.2). In particular, this should include possible changes in sources of raw materials and engineering modifications to preparation, filling and conveying equipment or to sterilisers, their equipment, instrumentation and services.

11.6   Records of these audits should be retained on file with the scheduled process establishment details.

11.7   Where the annual audit does not require confirmation of temperature distribution in the sterilisers, this should be confirmed every three years or whenever the validity of the scheduled heat process may be in doubt.

11.8   Even in the most efficient factories, "out of tolerance" conditions will occur. Where these affect a "critical" scheduled process parameter, e.g. filled weights, delay times, initial temperature, heat process conditions or steriliser operation, the goods concerned should be placed under "quarantine" pending further investigation to determine their disposal. Such goods should not be released for sale without authorisation from the Thermal Process Manager.

11.9   It is important to ensure that quarantined goods are clearly marked and segregated, and are not inadvertently shipped before their release has been properly decided and authorised. A demonstrably efficient quarantine system is important should a product recall become necessary.

12 ACIDIFIED (NORMALLY LOW ACID) FOODS

12.1 General Considerations

12.2 Process Requirements

12.3 Acidification

12.4 pH Measurement

12.1 GENERAL CONSIDERATIONS

12.1.1 Low-acid foods are considered to be those with pH values above 4.5. Within the United Kingdom this is the minimum accepted pH above which the growth of and toxin production by *Cl. botulinum* (the most heat resistant of pathogens) may be sustained. Acidification to pH 4.5 or below (by the addition of appropriate acidulent(s) or ingredients) may allow considerable reduction of the thermal process (less than F_o 3), to the benefit of product quality and without compromising product safety.

12.1.2 Acidification and reduced heat processing of naturally low-acid foods are very critical operations involving public health risks and appreciable losses of product if carried out incorrectly. It is essential that adequate acidification is achieved to reduce product pH to 4.5 or lower and that subsequent heat processing is sufficient to destroy organisms capable of outgrowth at the product pH achieved.

12.1.3 Acid tolerant sporeforming bacteria, yeasts and moulds may germinate and grow in acid conditions (pH 4.5 or lower). Heat resistance may be influenced by pH such that significantly more severe heat treatments are required for their destruction at pH 4.2–4.5 than at pH 3.8–4.2 or below pH 3.8.

12.1.4 The critical factors which influence the effectiveness of acidification and the thermal process requirements must be clearly understood and specified within the scheduled process documentation.

12.1.5   Examples of heat processed acidified low-acid foods include artichokes, cauliflowers, olives, peppers and some tropical fruits. Certain commodities, such as pears and tomatoes, have natural pH values in the region of 4.5 and the requirement to add acidulent needs to be carefully monitored, with addition as and when necessary.

12.2   PROCESS REQUIREMENTS

12.2.1   Localised rise in pH from growth of acid-tolerant organisms can allow outgrowth of non-acid-tolerant organisms surviving mild heat treatments. Cases have been known where growth of *Cl.botulinum* has occurred in this manner.

12.2.2   It is fundamental to successful preservation that the pH of the least acid component is regarded as that pH value which characterises acidified low-acid foods and therefore their heat process requirements. It follows that for heterogeneous foods, the pH value of a macerate cannot be accepted, in principle, as the critical factor, but it is used for formulation and control purposes.

12.2.3   Equilibrium pH throughout the product may not be achieved for a period of time, perhaps many hours or even days, is not easily controlled and may vary with different ingredients. The pH that characterises the acidified product must therefore be measured within a fixed time, e.g. 24 hours, of the end of the heat process. (But see also subsection 12.3.2 for process establishment criteria.)

12.2.4   Determination of the characteristic pH and scheduled heat process therefore requires careful and controlled evaluation during the development of the acidified low-acid product. Pack-to-pack variation in pH value is to be expected and therefore it is good practice to select a QC target value below that required for the scheduled process.

12.2.5   Scheduled processes for acidified low-acid foods should be established by suitably trained personnel with due regard to all the critical factors, including those relevant to the acidification process and the anticipated microbiological flora of the product. The heat treatment necessary to achieve commercial sterility will be specifically related to the pH value achieved and may be required to destroy acid-tolerant sporing bacteria, yeasts and moulds.

12.2.6   Critical factors which require specification within the GMP documentation for the acidification process would normally include, at least, the initial pH and nature of the foodstuff, the type and concentration of acidulent used, the means of addition and the method of mixing.

12.3   ACIDIFICATION

12.3.1   A number of substances, typically organic acids, are permitted additives for use in acidifying processes. The selection of acidulent(s) depends on a variety of factors, including flavour, cost, physical form, as well as effectiveness in achieving the required pH reduction.

12.3.2   The acidulent may be added to the food in a number of different ways, e.g. by addition to blanching water, by incorporation in a mix prior to filling, or by addition to individual containers. It is essential that the acidification process is controlled so that the pH of each part of the product is as specified and reliably at 4.5 or below within 4 hours of the end of the heat process.

12.3.3   Quality assurance procedures should be specified and control checks made to ensure that the required level of pH reduction is being achieved during the manufacturing process.

12.4   pH MEASUREMENT

12.4.1   pH measurement should be made using a properly calibrated instrument with a demonstrated accuracy of ±0.05 pH units. The pH range under consideration is small, and great care must be taken in calibration and standardisation of the instrument with appropriate buffer solutions, cleaning of electrodes, sample temperature control, etc. It is highly desirable that a second pH meter of sufficient accuracy be available in case of breakdown of the primary instrument. Records of pH meter calibration should be maintained.

12.4.2   pH electrodes used should be of a design appropriate to measure the pH of the individual components, since it is these values which are required rather than the pH value of an homogenised sample. The calibration of the electrodes should be checked at a frequency in accordance with the maker's instructions.

12.4.3   pH measurement at the end of the heat process should be made within the temperature range representing that expected in normal distribution and storage of the product.

13 SHELF-STABLE CANNED CURED MEATS

13.1 General Considerations

13.2 Mixing

13.3 Criteria for Thermal Processes

13.1 GENERAL CONSIDERATIONS

13.1.1 Shelf-stable canned cured meats are preserved partly by thermal processing but also by the introduction of salt and sodium nitrite in sufficient concentrations, and other factors, to provide a shelf-stable product. In relation to salt addition, the relevant controlling parameter is the percentage of salt within the aqueous phase.

13.1.2 The thermal processes used for these canned meats are relatively mild and allow the survival of significant numbers of bacterial spores. Outgrowth of these spores is prevented by chemical inhibition. (Corned beef is atypical in that the severity of thermal process required for satisfactory product textural characteristics achieves an F_O value considerably in excess of 3.).

13.1.3 The specification of the thermal process conditions must be made by suitably qualified personnel and must include all relevant critical factors in order to provide assurance of microbiological safety. F_O values are comparatively low and do not adequately express the effectiveness of heat treatment applied in achieving shelf stability. In defining a scheduled process, it is important to define not only sterilisation time and temperature but also the centre temperature attained by the product. The actual process temperature used may be as important as the F_O value attained. This is because F_O values relate to a z value of 10°C, and in cured meat products this particular z value may be inappropriate.

13.1.4 The primary factors affecting product safety are the levels of salt and sodium nitrite, pH, the initial microbiological quality of the food and the heat process applied. Additional factors that may affect safety and/or stability are available water, redox potential, and the addition of ascorbate/isoascorbate, nitrate and polyphosphates.

13.2   MIXING

13.2.1   The quantity of sodium nitrite to be added to a mix of meat is relatively small. In order to ensure adequacy of mixing, it is preferable to apply the sodium nitrite as part of a pre-mix diluted with common salt or other suitable ingredient.

13.2.2   The omission of sodium nitrite from a batch of meat, or substantial reduction in the input level below 150 ppm in the product mix, could have serious implications for product safety. It is vitally important, therefore, that a suitable control procedure is documented and employed within the factory in order to ensure that such omission does not happen.

13.3   CRITERIA FOR THERMAL PROCESSES

13.3.1   Typical combinations of brine concentrations and thermal processes, used in conjunction with an initial concentration of 150 ppm sodium nitrite in the meat mix, are recommended for products within the consultation paper prepared for the Codex Alimentarius Commission: ALINORM 86/16, Appendix VI.

13.3.2   Bulk incubation of significantly sized samples, with subsequent microbiological examination to determine whether growth has occurred, should be undertaken to provide substantiating evidence in the initial validation of each process used.

13.3.3   Any minimum safety criteria for the production of shelf-stable cured meats are likely to be inadequate if the meat ingredients are heavily contaminated with spores. As a general recommendation, the concentration of spores within the meat should not exceed 3/g of clostridial or 100/g of other mesophilic bacilli. For this reason, analysis of the meat supply and auditing of the suppliers should be included within the quality assurance procedures. The use of mechanically recovered meat may be accompanied by considerable increase in initial microbiological loading.

13.3.4   The mesophilic spore count for spices should not exceed 5 x 10^3/g and decontaminated spices should be used wherever possible.

13.3.5   The contribution of non-meat ingredients, other than spices, to the contamination of the final raw product should be collectively below 50 mesophilic spores/g.

13.3.6   **Pasteurised Meat Products**
It must be recognised that pasteurised meat products, typically produced in large Pullman cans, have been insufficiently processed to eliminate the spores of micro-organisms which may metabolise at ambient temperatures. They consequently require refrigerated storage subject to the Food Hygiene (Amendment) Regulations 1990 (these Regulations apply only to England and Wales), which specify that relevant food must be kept at or below 5°C (this requirement is under review).

The packaging of such pasteurised meat products must be clearly labelled with respect to the requirement for refrigerated storage.

14 COMMON INSTRUMENTS, CONTROLS AND FITTINGS FOR STERILISATION EQUIPMENT

14.1   MASTER TEMPERATURE INDICATOR (MTI)

Each retort or steriliser must be fitted with an independent MTI which may be either of the platinum resistance or mercury-in-glass type of suitable specification.

Platinum resistance thermometers should, as minimum, conform to the performance specification in CFDRA Technical Bulletin No. 61.

The mercury-in-glass thermometer should have a scale of not less than 152 mm (6 inches), which should be graduated at intervals of not more than 1 C degree (2 F degrees) over a range of about 50 C degrees (or 100 F degrees). It should be easily readable to 0.5 C degrees (1 F degree).

The sheath or tail piece length is an important feature. Since a long sheath results in a sensitive instrument but causes installation difficulties, a compromise must be sought, but 90 mm (or 3½ inches) may be regarded as a reasonable minimum. With steam retorts, in order to achieve a constant flow of steam over the sheath during the process, a 2 mm (1/16 inch) bleed should be drilled in a suitable position, remembering that during cooling some water will escape through this hole. Thermometers should be installed so that they can be read easily.

Black scale markings on a white background and the coloured mercury feature (red or blue) are useful in this respect. A scale range of 76°–130°C (170°–270°F) is usually considered acceptable.

Retort thermometers should be tested for accuracy against a KNOWN TRACEABLE standard on installation and at time intervals of not greater than six months. The calibration should be carried out in the sterilising medium and the thermometer placed in the operating position either within the steriliser or in a test rig. The check should be carried out at three individual temperatures within the normal processing temperature range. Dated records of the calibration check must be kept and, ideally, a dated tag attached to each instrument. The records should show the standard used, method used and personnel carrying out the test. Thermometers with errors greater than 0.5°C from standard should be taken out of service and repaired or destroyed.

14.2   TEMPERATURE CONTROLLER

Each retort or steriliser must be fitted with a temperature controller which is capable of controlling the sterilising environment temperature to –0.5 to +1.0°C (–1 to +2°F). The instrument is often of the recording/controlling type. (See subsection 10.3.4.)

14.3   TEMPERATURE/TIME RECORDING DEVICES

Each retort or steriliser must be equipped with at least a temperature/time recording instrument which provides a permanent record of each sterilisation cycle. This instrument may be part of a single unit with the temperature controller and must be calibrated at least once per year.

The recorder should agree with the MTI to within 0.5°C and should not read higher than the MTI at the scheduled sterilising temperature.

Where paper charts are used, it is essential that the correct chart is used for each recorder. Each chart should have a working scale of not more than 12°C per cm (55°F per inch) and the accuracy of the instrument should be at least ±0.5°C (1°F). Where appropriate, the chart should be aligned with the real time.

During operation, the recorder should be checked for accuracy against the MTI at least hourly for continuous cookers and during each cycle for batch retorts.

Where an external well is used to house the recorder sensing point at least a 2.0 mm (1/16 inch) bleed should be present which emits steam or water continuously during the sterilising period.

The recorder adjustment control should normally be kept locked so that unauthorised alterations cannot be made. Any adjustments made should be recorded.

14.4 PRESSURE GAUGE

Each retort or steriliser must be fitted with a pressure gauge. The gauge should have a face of at least 102 mm (4 inch) in diameter and a scale graduated in divisions of not greater than 0.14 kg/sq cm (2.0 psi). The gauge should have an accuracy of ±1% of full scale deflection. Ideally, the gauge should have a range from zero such that the safe working pressure of the retort or steriliser is about two-thirds of the full scale. The gauge should be checked for accuracy at least once per year.

14.5 BLEEDS

All retorts and sterilising systems using saturated steam as a sterilising medium must be fitted with at least one bleed located at the opposite end of the vessel to the steam inlet. The bleeds must be fully open during any sterilisation operation. Such bleeds must be at least 3 mm (1/8 inch) in diameter and be so positioned that the operator can easily see that they are functioning correctly, i.e. steam is issuing.

14.6 TIMING DEVICES

When a sterilisation cycle is under manual control, one accurate, easily read timer should be visible to the operator for each batch of retorts when carrying out operations and recording data. The timer should be calibrated in one minute divisions.

15 TYPES OF STERILISATION EQUIPMENT

15.1   Vertical Steam Retorts
15.2   Vertical Water Retorts
15.3   Horizontal Steam Retorts
15.4   Horizontal Steam/Air Retorts
15.5   Horizontal Showered Water Retorts
15.6   Horizontal Total Immersion Water Retorts
15.7   Crateless Retorts
15.8   Hydrostatic Sterilisers
15.9   Reel and Spiral Sterilisers

15.1   VERTICAL STEAM RETORTS

Heating is within an atmosphere of saturated steam. The total removal of air is vital in order to avoid cold spots. Temperature is controlled and pressure is defined as a consequence.

15.1.1   **Master Temperature Indicators**
See subsection 14.1.

15.1.2   **Temperature Controllers**
See subsection 14.2.

15.1.3   **Temperature/Time Recording Devices**
See subsection 14.3.

15.1.4   **Pressure Gauges**
See subsection 14.4.

15.1.5   **Bleeds and Condensate Removal**
See subsection 14.5.

Means must be provided for the removal of condensate from the base of the retort during sterilisation. In simple systems this may involve operation with the drain valve cracked open; or alternatively, in more sophisticated systems more complex automatic condensate removal equipment may be deployed. It is important, however, that the operator is able to observe that the system is functioning properly as build-up of condensate could cause localised understerilisation.

15.1.6 **Timing Devices**
See subsection 14.6.

15.1.7 **Steam Inlet**
The steam inlet to each retort should be large enough to provide sufficient steam for proper operation of the retort and should enter at a suitable point to facilitate air removal during venting.

15.1.8 **Crate Supports**
A bottom crate support shall be designed and used so as not to affect venting. Baffle plates should not be used in the bottom of retorts. Centring guides should be installed to ensure clearance between the retort crate and the retort wall.

15.1.9 **Steam Spreaders**
The perforated steam spreaders, if used, should be in the form of a cross or coil. The number of perforations should be such that the cross-sectional area of the perforations is equal to one and a half to two times the cross-sectional area of the smallest restriction in the steam inlet line.

15.1.10 **Stacking Equipment and Position of Containers**
Crates, trays and layer pads should be so constructed to aid venting and good temperature distribution. When perforated sheet metal is used for the bottoms, the perforations should be equivalent to 25 mm (1 inch) holes on 50 mm (2 inch) centres. If dividers are used between the layers of containers, they should be perforated as above. To avoid the masking of perforations, the dividers must only be used singly.

A layer pad should never be placed over a retort crate base as misalignment of the holes may prevent adequate steam circulation. Care must be taken that any unperforated margin of a layer pad does not mask the base of any can size which may be used.

15.1.11   **Vents**
           Vents should be located at the opposite end of the retort
           to the steam inlet. Vents must not be connected directly to
           a closed drain system without an atmospheric break in
           the line. Where a retort manifold is used, the manifold
           shall be sized so that the cross-sectional area of the pipe
           is larger than the total cross-sectional area of all
           connecting pipes. The manifold discharge pipe should
           lead directly to the atmosphere.

15.1.12   **Air**
           Retorts using air for pressure cooling must be equipped
           with a suitable valve to prevent any air leakage into the
           retort during processing. The use of two valves in series or
           duplex valves should be considered, although this is not
           regarded as a substitute for good maintenance.

15.1.13   **Water Valves**
           It is equally important that retorts using water for
           cooling must be equipped with a suitable valve to prevent
           leakage of water into the retort during sterilisation,
           otherwise localised reduction of temperature will occur,
           with the possible consequence of understerilisation.

   15.2   VERTICAL WATER RETORTS

           Sterilisation is within an environment of superheated
           water. Air overpressure is necessary to prevent water
           evaporation and container deformation and to aid
           retention of lids of glass jars. Temperature and pressure
           are independently controlled.

15.2.1   **Master Temperature Indicator**
           See subsection 14.1.

           The MTI instruments should be positioned in the
           thermometer pocket approximately half way up the side
           of the retort and such that they are beneath the surface
           of the water throughout the sterilisation cycle.

15.2.2   **Temperature Controller**
           See subsection 14.2.

15.2.3   **Temperature/Time Recording Devices**
           See subsections 14.3 and 15.2.1.

15.2.4   **Pressure Gauge**
See subsection 14.4.

15.2.5   **Timing Devices**
See subsection 14.6.

15.2.6   **Steam Inlet**
The steam inlet should be large enough to provide sufficient steam to ensure a good temperature distribution within the retort.

15.2.7   **Crate Supports**
See subsection 15.1.8.

Centring guides should be installed so as to ensure there is 35 mm (1.5 inch) clearance between the side wall of the crate and the retort wall.

15.2.8   **Steam Spreaders**
The steam should be distributed in the bottom of the retort in a manner adequate to provide uniform heat distribution throughout the retort.

15.2.9   **Drain Valve**
A non-clogging, watertight valve should be used.

15.2.10   **Water Level Indicator**
The containers must always be covered by process water. The height of the processing water must, therefore, be visible in a sight glass or shown by means of an accurate automatic level control/detector.

THE CONTAINERS MUST BE COVERED BY AT LEAST 10 CM (4 INCHES) OF PROCESS WATER.

15.2.11   **Water Circulation**
Water agitation or circulation systems, whether operated by pumps or compressed air, should be installed and used in such a manner that a specified and reliable temperature distribution is maintained throughout the retort.

The pump should be equipped with an alarm system to indicate malfunction of water circulation for whatever reason. (Note that an alarm which indicates power failure to the motor does not do this. It has been known for a pump to fail even though the shaft is turning.).

Modification to the water circulation method used in the retort must be confirmed by heat distribution studies showing uniform temperatures within the retort.

Crates and trays etc. for containers should be designed so that the circulation of the water is sufficiently unimpaired so that temperature distribution throughout the retort is within specification.

15.2.12 **Air Supply and Control**
If air is used to promote circulation, it should be introduced into the steam line at a point between the retort and the steam control valve at the bottom of the retort.

The retort pressure should be controlled by an automatic pressure control unit.

15.3 HORIZONTAL STEAM RETORTS

Heating is within an atmosphere of saturated steam. The total removal of air is vital in order to avoid cold spots. Temperature is controlled and pressure is automatically defined in consequence.

15.3.1 **Master Temperature Indicators**
See subsection 14.1.

15.3.2 **Temperature Controllers**
See subsection 14.2.

15.3.3 **Temperature/Time Recording Devices**
See subsection 14.3.

15.3.4 **Pressure Gauge**
See subsection 14.4.

15.3.5 **Timing Devices**
See subsection 14.6.

15.3.6 **Steam Inlet**
See subsection 15.1.7.

15.3.7   **Crate Supports**
          See subsection 15.1.8.

15.3.8   **Steam Spreaders**
          Steam spreaders should normally be perforated pipes,
          the length of the retort. Retorts over 10 metres (30 feet)
          long should have at least two steam inlets connected to
          the spreader at approximate equal divisions of its
          length. The number of perforations should be such that
          the total cross-sectional area of the perforations is equal
          to one and a half to two times the cross-sectional area of
          the smallest restriction in the steam inlet line.

15.3.9   **Bleeds**
          See subsection 14.5.

          Ideally, bleeds should be located approximately 300 mm
          (1 foot) from each end along the top of the retort and
          additional bleeds should be located not more than 2400
          mm (8 feet) apart along the top. Bleeds may be installed
          at positions other than those specified above, so long as
          there is reliable heat distribution data to indicate
          adequate air removal during venting.

          A bleed should be installed below the bottom layer of
          containers to remove condensate during the sterilising
          period. (Simple systems may operate with the drain
          valve cracked open.)

15.3.10   **Stacking Equipment and Position of Container**
          See subsection 15.1.10.

15.3.11   **Vents**
          See subsection 15.1.11.

          Other vent pipe arrangements, which differ from that
          described in subsection 15.1.11, may be used provided that
          there is direct evidence that they consistently
          accomplish adequate venting.

15.3.12   **Air Valves**
          See subsection 15.1.12.

15.3.13   **Water Valves**
          See subsection 15.1.13.

15.4 HORIZONTAL STEAM/AIR RETORTS

The retorts are designed to use an homogenous mixture of steam and air as the heat transfer medium. Temperature and pressure are independently controlled and the total removal of air during a venting procedure is therefore not necessary.

15.4.1 Master Temperature Indicators
See subsection 14.1.

The thermometer should be positioned such that its bulb of sensing point is fully in the path of the circulating steam/air mixture.

15.4.2 Temperature Controller
See subsections 14.2 and 15.4.1.

15.4.3 Temperature/Time Recording Devices
See subsections 14.3 and 15.4.1.

15.4.4 Pressure Gauge
See subsection 14.4.

15.4.5 Condensate Removal
Provision must be made at the bottom of the retort for the removal of condensate during the sterilisation cycle. This will ensure that the bottom containers do not become submerged in water. Ideally, an automatic valve triggered by a level probe should be used.

15.4.6 Timing Devices
See subsection 14.6.

15.4.7 Steam Distribution
The steam inlet should run along the whole length of the retort with a series of holes sized and spaced to give an even distribution of steam. See also subsection 15.3.8.

15.4.8 Fan/Circulation System
All retorts of this type must be fitted with a fan or other method of circulating the steam/air mixture. This is to prevent the formation of low temperature pockets and to ensure the desired rate of heat transfer. The circulation system must provide an acceptable temperature distribution in the retort, and no thermal process should be scheduled until confirmed by careful testing.

A method must be provided whereby the operation of the fan may be checked during the sterilisation cycle. A sensor, such as a proximity detector, may be used but the verification of operation must be on the fan side of any drive coupling. Failure of the fan for any reason must be detected and should activate an audible or audio-visual alarm.

15.4.9   **Crate/Separator Design**
Where crates or frames are used to retain containers during the retort cycle or separators are used between layers of containers, they should be designed to allow circulation of the steam/air mixture which is sufficiently unimpaired so that the temperature distribution throughout the retort is within specification.

Where low profile containers are to be sterilised, it may be necessary to make special provisions for the circulation of the sterilising medium. In the case of retort pouches, there may be a need to restrain the allowable expansion of the pouch by means of the separator.

15.4.10   **Overpressure Control**
A pressure control system should be available which controls the operation of both the air inlet and steam/air outlet valves. A continuous record of the retort pressure should be kept using a chart recorder.

Control of the overpressure will allow the internal pressure in the container that is generated by gas expansion to be balanced. Container volume can be controlled by the overpressure applied. This is particularly important with flexible containers.

15.4.11   **Rotation/Agitation**
See subsection 8.5.5.

Where rotation or agitation of the containers in the retort is part of the scheduled process, the speed of rotation should be regularly checked. Ideally, a recording tachometer should be used to log this data. Rotation failure should activate an appropriate alarm system. Power supply to the drive motor is not sufficient evidence that correct rotation is occurring.

NB Container headspace control is a critical factor in rotary processes heating by forced convection.

15.5 HORIZONTAL SHOWERED WATER RETORTS

These retorts use a low volume of superheated water circulated at a high flowrate through specifically designed shower heads to sterilise containers. As overpressure is required to ensure that the water remains liquid, a venting procedure is not required. Temperature and pressure are independently controlled. Overpressure is applied by the introduction of compressed air.

15.5.1 Master Temperature Indicators
See subsection 14.1.

The thermometer shall be positioned such that its bulb or sensing element is fully immersed in the flow of the circulating water on the outlet side of the vessel.

15.5.2 Temperature Controller
See subsections 14.2 and 15.5.1.

Where the heating and cooling of the vessel is indirect, the sensing probe for the temperature controller is normally sited immediately after the heat exchanger.

15.5.3 Temperature/Time Recording Devices
See subsections 14.3 and 15.5.1.

In vessels that are heated indirectly, it is advisable to site the sensing probe for the temperature/time recording device in the return pipe to the heat exchanger.

15.5.4 Pressure Gauge
See subsection 14.4.

15.5.5 Timing Devices
See subsection 14.6.

15.5.6 Circulation Pump
In retorts of this type, it is essential that any failure of the water circulation pump activates an audible alarm system and that the countdown timer for the sterilisation cycle is stopped. This may be achieved by measurement of pressure at the pump dischange.

15.5.7   **Shower Head**
It is essential that the water flow through the shower head is unimpaired. The state of the holes in the shower head must be checked at least weekly to ensure that they are not becoming blocked due to scale deposition or product from damaged containers. If the holes show signs of blockage, the shower head must be cleaned either manually or by circulation of a solution of a suitably inhibited acid cleaner in accordance with the retort manufacturer's instructions.

Where orifice plates are used to control the distribution of water, they should be checked at least twice per year for fouling. Where maintenance work involves dismantling of pipework etc, care must be taken to refit any orifice plates present in the correct position.

15.5.8   **Suction Inlet/Pump Filter**
Design of the area around the suction inlet of the circulation pump should be such that it is not possible for the inlet to be blocked by fallen containers. A filter should be fitted upstream of the circulation pump. This filter should be removed and cleaned at least weekly.

15.5.9   **Refilling**
Where a showered water retort is topped up with water at the start of a sterilisation cycle, further additions of water during the cycle must be prevented to stop the possibility of post-process contamination.

15.5.10   **Overpressure**
See subsection 15.4.10.

It should be noted that an air overpressure is always required to keep water liquid at temperatures over 100°C. Manufacturer's literature and steam tables should be consulted before any sterilisation cycle is accepted as part of the scheduled process.

15.5.11   **Rotation/Agitation**
See subsection 15.4.11.

15.5.12   **Heat Exchangers**
Where heating and cooling of the process water is indirect via a heat exchanger it is advisable to clean

both sides of the exchanger with corrosion inhibited acid to maintain the highest rates of heat transfer.

15.5.13 Cooling Water

Where heating and cooling of the process water is direct, only water of the specified microbiological quality may be used for cooling.

15.5.14 Crate/Separator Design

Where crates, frames or separators are used between layers of containers, the holes therein must be sized and spaced to allow unimpaired flow of water over the containers. In shower-type retort systems where water flow is from the top to the bottom of the vessel, retort crates often have unperforated sides.

15.6 HORIZONTAL TOTAL IMMERSION WATER RETORTS

These retorts use superheated water as the heat transfer medium and typically comprise two pressure vessels. The upper vessel is a pre-heating boiler and the lower vessel is used for sterilisation. Product containers are fully immersed during sterilisation and the water is usually circulated by pump. Overpressure is usually achieved with an independent steam cushion acting on the surface of the water in the upper vessel. Pressure and temperature are independently controlled.

15.6.1 Master Temperature Indicators

See subsections 14.4 and 15.2.1.

The position of the MTI should always be on the centre line of the retort. Minimum depth of insertion should be 5 cm (2 inch).

15.6.2 Temperature Controller

See subsections 14.2 and 15.6.1.

15.6.3 Temperature/Time Recording Devices

See subsections 14.3 and 15.6.1.

15.6.4 Pressure Gauge

See subsection 14.4.

15.6.5 Timing Devices

See subsection 14.6.

15.6.6	**Steam Inlet**

See subsection 15.2.6.

15.6.7	**Water Heating**

Water in the pre-heater is heated by the introduction of steam through a spreader pipe in the base of the vessel. During sterilisation, however, live steam is introduced into the circulation system of the lower vessel in order to maintain the scheduled temperature.

15.6.8	**Water Level Indicator**

See subsection 15.2.10.

15.6.9	**Water Circulation**

See subsection 15.2.11.

15.6.10	**Air Supply and Control**

See subsection 15.2.12.

15.6.11	**Overpressure Control**

Sufficient overpressure must be applied to maintain water in the liquid phase at the processing temperature. The overpressure may be applied by use of compressed air, or in two vessel systems more usually by steam.
Overpressure may also be used to control the expansion of containers due to the expansion of gas within the product.

A continuous record of the retort pressure should be kept using a chart recorder.

15.6.12	**Crate/Separator Design**

Where crates or frames are used to retain containers during the retort cycle or separators are used between layers of containers, they should be designed to allow circulation of water which is sufficiently unimpaired so that the temperature distribution throughout the retort is within specification.

In retorts where the hot water storage vessel volume is less than that of the retort, crates of dummy product will be required to fill the retort volume if part loads are to be sterilised.

Where the volume of hot water stored is equal to or greater than the volume of the retort, the temperature distribution in the retort should be checked with part loads before this practice is accepted.

15.6.13 **Rotation/Agitation**
See subsection 8.5.5.

15.7 CRATELESS RETORTS

These retorts are large pressure vessels which are top loaded with cans when the retort is full of water. The water acts as a cushion to break the can's fall. Water is purged from the retort with steam prior to the venting and steam sterilisation stages. Cans are pressure cooled with chlorinated water and discharged into a chlorinated cooling water canal.

5.7.1 **Master Temperature Indicators**
See subsection 14.1.

15.7.2 **Temperature Controller**
See subsection 14.2.

15.7.3 **Time/Temperature Recording Devices**
See subsection 14.3.

15.7.4 **Pressure Gauge**
See subsection 14.4.

15.7.5 **Bleeds and Condensate Removal**
See subsections 14.5 and 15.4.5.

Any condensate removal device should be sited below the lowest level of containers in the sterilising vessel. Bleeds should not discharge into the cooling canal.

15.7.6 **Timing Devices**
See subsection 14.6.

15.7.7 **Steam Supply**
As steam is supplied via ports at the top of the retort, rather than through the conventional spreader pipe arrangement, care must be taken to optimise steam supply parameters (pressure and volume flow) at the commissioning stage.

15.7.8   **Vents**

See subsection 15.1.11. Other vent arrangements which differ from that described in subsection 15.1.11 may be needed for adequate venting of this type of sterilising vessel. Evidence that these arrangements accomplish adequate venting should be ideally gathered by experiment.

15.7.9   **Air Valves**

See subsection 15.1.12.

15.7.10   **Water Valves**

See subsection 15.1.13.

15.7.11   **Discharge Door Baffling**

To prevent jamming of cans on discharge, it is necessary to fit baffles internally above the discharge door. It is important that these baffles do not impair the free circulation of steam in the retort during processing. The area beneath the baffles must be investigated with suitable temperature measuring equipment during commissioning.

15.7.12   **Can Counters/Ink Jet Coders**

These devices are desirable prior to retort loading to ensure traceability of all cans through the system. A quality audit should be regularly applied to these systems.

15.7.13   **Fill Weight Control**

It is advisable to have check-weighers on line to reject under-weights and prevent potential floating containers from entering the processing vessel.

15.7.14   **Container Loading**

It is essential that the discharge door on the processing vessel is interlocked with the loading door and loading conveyor so that it is not possible for cans to pass through the vessel unprocessed.

15.8   HYDROSTATIC STERILISERS

In hydrostatic sterilisers, the pressure within the sterilising chamber is balanced on either side by hydrostatic legs of water. Cans are conveyed continuously through the system in carrier bars driven by a continuous chain. The sterilising medium is usually saturated steam, so that if temperature is the controlled variable, pressure is automatically defined. In some systems, primary control is effected by pressure measurement and temperature automatically defined in consequence. Hot water may also be used as the designed heat transfer medium when similar technological factors to those for horizontal showered water retorts apply (see subsection 15.5).

15.8.1   **Master Temperature Indicators**
See subsection 14.1.

At least two MTIs are to be fitted in each steam dome, one located just above the normal steam/water interface and the other at the top of the dome. Where the scheduled process specifies the maintenance of given temperatures in the hydrostatic legs, at least one MTI must also be installed in each leg.

15.8.2   **Temperature Controller**
See subsection 14.2.

15.8.3   **Temperature/Time Recording Devices**
See subsection 14.3.

At least two recorder probes are to be fitted in each steam dome, one located just above the steam/water interface and the other at the top of the dome. Where the scheduled process specifies the maintenance of given temperatures in the hydrostatic legs, recorder probes must be installed at the top and bottom of each leg. Additional probes may be required for areas that are found to be critical.

15.8.4   **Pressure Gauges**
See subsection 14.4.

15.8.5   **Bleeds and Condensate Removal**
See subsection 14.5.

An alternative method of bleeding may be provided via the manual vent valves located immediately above the steam/water interface of the steriliser. These valves are left open to remove air during part of the bringing-up procedure. During normal operation these valves are closed but 3 mm (1/8 inch) diameter holes may be provided in the valve to continuously bleed the steam dome.

Where the operator is located away from the steriliser, a method of detecting the operation of the bleed must be provided.

Where the design of the hydrostatic steriliser gives rise to a steam chamber remote from the hydrostatic legs, a separate condensate drain must be provided.

15.8.6   **Timing Devices**
See subsection 14.6.

15.8.7   **Vents**
See subsections 15.1.11, 15.8.5 and 15.8.8.

15.8.8   **Bringing Up**
The steriliser should only be brought up to operating temperature by following an approved method which has been specified by either the manufacturer or canner. The method must ensure that adequate venting occurs to remove air from the steam dome such that the steriliser achieves the minimum target of ±0.5°C of the specified sterilising temperature.

15.8.9   **Water Level Control**
An automatic system must be provided to control the level of water continuously at the bottom of the steam dome and ensure that the conveyor chain does not run in water.

15.8.10   **Conveyor Speed**
The sterilisation time in hydrostatic sterilisers is determined by the speed of the conveyor chain through the steam dome. The conveyor chain speed must form part of the scheduled process and must be measured at the start of processing and at intervals of, at most, 2 hours if a manual check is performed. Where the conveyor speed is measured automatically, this should be displayed and recorded continuously.

While the number of container carriers in the steam dome is specified by the manufacturer of the hydrostatic steriliser, the actual number can vary as the height of the steam/water fluctuates. It is recommended that if the conveyor chain speed is less than 10 carriers per minute, the high level alarm position of the steam/water interface should be used as the reference point for the start of the steam dome.

An automatic system should be available to stop the steriliser conveyor if the temperature in the steam dome drops by more than 1°C (1.5°F) below the sterilising temperature. Such a temperature drop should be immediately alarmed to the steriliser operator. A means of preventing unauthorised conveyor speed changes should be provided.

15.8.11 **Hydrostatic Legs**
Where the temperatures in the hydrostatic legs are specified as part of the scheduled process, the temperatures must be controlled to ±5°C (9°F) of the specified temperature.

The practice of transferring water from one leg to the other for energy conservation should be reviewed. Water from the pre-heat leg, which may be dirty and microbiologically contaminated, should never be transferred to the second leg. The microbiological status of the water legs should be checked periodically.

The water in the hydrostatic legs will be pasteurised during the bringing-up procedure. However, where the normal operating conditions are less than 80°C, consideration should be given to regular pasteurisation of the pre-cool leg water.

In hydrostatic sterilisers where there is no pumped circulation of the pre-cool leg water, cooling water should not be added to control the leg temperature.

15.8.12 **Schematic Drawing**
Ideally, a schematic drawing of the steriliser should be available at the operator station. This drawing should clearly show the number of container carriers in each section of the steriliser to aid the isolation of product in the event of processing deviations.

15.9 REEL AND SPIRAL STERILISERS

These are cylindrical process vessels operating in series and connected together in a modular manner. There will always be a sterilising vessel and a cooling vessel. Sterilisation is in saturated steam or superheated water. Cans enter the system through a rotary pressure valve and effectively roll, slide and carry continuously through the system.

15.9.1   **Master Temperature Indicators**
See subsection 14.1.

15.9.2   **Temperature Controllers**
See subsection 14.2.

15.9.3   **Temperature/Time Recording Devices**
See subsection 14.3.

15.9.4   **Pressure Gauges**
See subsection 14.4.

15.9.5   **Timing Devices**
See subsection 14.6.

15.9.6   **Steam Inlet**
See subsection 15.3.8.

15.9.7   **Bleeds**
See subsections 14.5 and 15.3.9.

15.9.8   **Venting and Condensate Removal**
Bringing these large process vessels to operating temperatures is an operation requiring considerable care in order to minimize mechanical stress. Manufacturers' instructions should be carefully followed. Vents should be located in that portion of the steriliser opposite to the steam inlets, and the venting procedure employed should have been experimentally validated and appropriate records kept on file.

Provision is required for continuous drainage of condensate from the steriliser during the sterilising operation. The condensate bleed in the bottom of the shell serves as an indicator of condensate removal. It is useful to install a high condensate level alarm probe at each end of and at the bottom of the retort steam shell.

15.9.9 Steriliser Rotation

The rotational speed of the steriliser effectively determines the sterilising time and must therefore be specified in the scheduled process. The speed must be adjusted when the steriliser is started, whenever speed change is made, and at intervals of sufficient frequency so as to ensure that the steriliser speed is maintained as specified in the scheduled process. The speed should be checked against a stopwatch at least every four hours. Where the steriliser speed is measured automatically, this should be displayed and recorded continuously.

A means of preventing unauthorised speed changes on sterilisers should be provided.

16 COOK ROOM OPERATIONS

16.1 General Requirements

16.2 Separation and Identification of Pre- and Post-process

16.3 Process Status Indication and Batch Identification

16.4 Retort Operation Records

16.5 Control Checks

16.6 Heat Process Deviations

16.1 GENERAL REQUIREMENTS

16.1.1 The basis for sound cook room operations is that the equipment should be well designed, serviced, instrumented and maintained. Pipework and valves should be clearly distinguishable, easy to reach and with indicators clearly visible from the operating position. The operators should be well trained and effectively supervised, and documented operating and check procedures should be displayed or to hand.

16.1.2 It is essential that operators are effectively trained on the plant that they will be operating. Background information should also be given on microbiology, food hygiene and the principles of canning, including leaker infection, but care must be taken not to over-complicate the presentations whilst giving sufficient information to make the operations meaningful. In addition, training must be given on the check systems and record keeping, making it clear to the operator why such checks and records are necessary.

16.1.3 The requirements for the training of cook room supervisors are similar to those for operators. However, supervisors should be given additional information on the principles of sterilisation and on the interaction of the cook room operations with the rest of the factory departments. Further information should also be given on the sterilisers, the control systems and instruments, quarantine procedures, and on procedures to minimize post-process contamination.

16.1.4   In order to supervise the cook room operation effectively, the supervisor should not only be knowledgeable and diligent but should also have the opportunity to devote such time as is necessary to cook room supervision. This means that he should not be given duties which conflict with his primary responsibility which is to ensure sound cook room operation.

16.2   SEPARATION AND IDENTIFICATION OF PRE- AND POST-PROCESS

16.2.1   The cook room is the dividing line between the microbiologically contaminated raw material handling area and the clean post-process area. It is desirable that the spread of infection into the cook room be minimized and that adequate precautions are taken to prevent contamination of processed containers from unprocessed containers, crates and conveying equipment. It is important that processed containers are not manually handled whilst wet (see subsection 17.7).

16.2.2   In container handling systems for continuous sterilisers and fully automated batch sterilisers, good engineering should provide for the complete segregation of unsterilised and sterilised containers, either by physical or spatial barriers. Care should be taken with the detailed layout since, on very rare occasions, malfunction of the conveying system has resulted in containers "jumping" from one runway to another. This may result in incorrect processing or labelling and may also happen if containers become "hung up" or lodged on conveyor systems, especially at turning points on dead plates or where multilayer slat conveyors divide or join.

When, in an emergency, unsterilised or sterilised containers have to be discharged from the system, they should be collected in baskets or other suitable receptacles, identified clearly and prominently with their process status, and segregated until a decision has been made on their disposal.

16.2.3   Containers occasionally fall from baskets or jump from runways and may come to rest some distance from their original starting position. As there is usually no way of being absolutely certain of the status of such containers, they *must* be destroyed. To reinforce this requirement, prominent notices should be posted in all areas where this is a requirement.

It is also essential to ensure that the lines are clear of containers before the start and after the end of production or when a change of product (variety) occurs.

16.2.4   With batch sterilisers, one means of separation is the use of horizontal retorts with front entry, rear exit doors. If such equipment is not available, wherever possible a one-way system should be used, unprocessed cans entering the area at one side and processed cans leaving at the opposite side. Crates of processed or unprocessed cans should only be allowed to accumulate in specified sections of the retort area where segregation and prevention of contamination can be ensured. Particular care is required with vertical retorts to ensure that when crates of cans are passed over retorts, drips of contaminated water do not fall onto processed cans in open retorts.

16.3   PROCESS STATUS INDICATION AND BATCH IDENTIFICATION

16.3.1   Continuous and fully automated batch loading and unloading systems should ensure that, unless there is a malfunction, the correct scheduled process is always given.

A documented procedure is required for setting up the lines and for checking that all cans are given the correct scheduled process. All automated systems must be designed to be fail-safe.

Specific precautions must be built into the lines to prevent cans falling off the lines in the event of sudden stops or jams.

Procedures must be available for batch identification on continuous and batch lines so that when product changes or out-of-tolerance conditions occur (such as fill variations or excessive pre-process delay times), product may be suitably separated.

16.3.2   Manual batch operations are always open to error. It is therefore important that control and check systems are in use to minimize the possibilities of errors and to ensure that any errors will be detected and reported and appropriate corrective action taken.

16.3.3　It is important that the product inside a container and its process status are easily identifiable to both cook room and post-process personnel. Baskets of cans which require different scheduled processes must be distinguishable at a glance.

16.3.4　Multiple check systems are advisable but must be simple to ensure that they can be properly applied and are fully effective. Many systems can be used to achieve the requirements given in subsection 16.3.3. The following is an example of a system known to work well in practice. The basic principles are:

a. A heat sensitive indicator marked with the basket number and sufficient information to fully identify that basket of product;
b. A large, easily visible marker to identify at a glance the process status and product scheduled process group.

A typical system uses a heat sensitive indicator card or tape which is marked by the basket loader with the sequential basket number, date, product/line code, plus any other critical information such as time of starting to fill basket. In addition, a process status plate may be hung on the retort basket. This is a large (e.g. 100 cm^2) plate coloured red for unsterilised cans, and of a different shape for each line or scheduled process group.

16.3.5　The retort operator checks the information on the indicator and the can code and size, marks the indicator with the retort number, and records the basket numbers on his log, together with all other relevant information.

16.3.6　At the end of the heat process cycle, the retort operator checks the recorder trace and log to ensure that no process deviations have occurred and then removes the baskets from the retort. As each basket is removed, the heat sensitive indicator is checked for correct colour change, and the red process status plate is exchanged for an identical but green sterilised status plate. Should the heat sensitive indicator not show the correct colour change or it is lost, the operator must add a status deviation note marked with the retort number and cycle and inform his supervisor who must check and agree the status of that basket of product. The supervisor initials the status deviation note and, if it is agreed that the crate has been processed, adds the green status plate.

16.3.7   Baskets of cans must not be unloaded unless they carry the green status plate, a heat sensitive indicator showing the correct colour change and appropriate batch markings or the countersigned process status deviation note. A prominent notice to this effect should be displayed in the unloading area.

16.3.8   The heat sensitive indicators should be retained and sufficient records made to enable each pallet load of final product to be traced back to specific basket numbers and retort processes.

NB Heat sensitive indicators do not show that the CORRECT heat process has been used; this must be confirmed from the retort operator's log and recorder charts.

16.3.9   At the end of a production period, the independent records of numbers of baskets filled, numbers sterilised and numbers unloaded should be reconciled and checked by a designated person. This should be done not later than the next working day.

16.4   RETORT OPERATION RECORDS

16.4.1   It is important that the scheduled heat process is correctly applied, supervised and documented to provide positive assurance that all the requirements have been met. All records must be permanent, legible and correctly dated, and must be retained on file for a minimum of three years or for the shelf-life of the product, if this is longer.

16.4.2   The records must be accurate and should be recorded on a prescribed document, at the time the observation takes place, by the retort operator or other designated person. It must be possible by analysis of records to show that critical factors have been met and that cross-checks have been carried out at appropriate frequencies. Irrespective of whether the system is batch, continuous, manual or fully automatic, A FORMAL MANUALLY COMPLETED RETORT LOG IS REQUIRED.

For batch retorts, the log should include product and container identification, batch code, retort identification, date and time of observations, critical product factors which may be checked (e.g. initial temperature, delay times) and instrument readings as appropriate to the particular retorts in use. Each record should contain space for comments and should be signed by the person who fills in the log. Records should also be kept of critical operational conditions or instrument readings during the period when the retort is brought up to operating temperature.

16.5   CONTROL CHECKS

16.5.1   Independent (e.g. QC) checks should be carried out at least twice per shift on instrument readings, log sheets and recorder charts to ensure that the specified thermal processes are being properly achieved.

16.5.2   The Thermal Process Manager or his designate should check and sign all retort logs, recorder charts and other documentation relevant to the heat process before any product is released for sale.

16.6   HEAT PROCESS DEVIATIONS

16.6.1   Whenever checks, records or other means show that product may have received a heat process which is less severe than the scheduled heat process, the affected product must be identified, isolated and placed in quarantine pending investigation by the Thermal Process Manager or his designate. The identification of a process deviation requires specification, both in operator training and in the provision of written instructions, which must be permanently available in the retort area. It is also sound practice to have detailed written instructions available to the operators and supervisors on the details of how to proceed when a process deviation is discovered; this should include the personnel to be informed and the records to be made. When a deviation results in overprocessing, it is normal practice to quarantine the product for sensory analysis.

16.6.2   In certain circumstances, where a process deviation is discovered whilst the product is still undergoing heat treatment, it may be possible to modify the process to compensate. If such procedures are to be used, they must be fully documented and affected stock should be placed in quarantine to enable full investigation by the Thermal Process Manager.

16.6.3   Any low acid product which has not received a botulinum cook must be fully reprocessed to render it commercially sterile or destroyed under adequate direct supervision to ensure protection of the public health.

NB Heat processing times for reprocessed products may be very different from the scheduled heat process. This is because the partial process may have significantly altered the heat transfer characteristics (e.g. viscosity) of the product.

16.6.4   Full records should be made for any product affected by a process deviation, showing the evaluation procedures used, the results obtained and the actions taken.

17 POST-THERMAL PROCESS PROCEDURES

17.1   INTRODUCTION

It is necessary to minimize the incidence of microbial post-thermal process re-infection in order to avoid not only the financial and economic consequences but also the possible public health risk of infected non-blown cans being unsuspectingly offered for sale. The design and manufacture of the open-top type of food container has reached a state of high technical development. However, it is known that for a short period of time following the sterilising process, the integrity of the hermetic seal is vulnerable and microleakage may occur. Satisfactory post-thermal process hygiene measures must therefore be adopted and maintained.

17.2   CONTAINER COOLING

17.2.1   Water used for container cooling purposes should be of low bacterial content. (See under Water Supply, subsections 5.12 and 17.3.)

17.2.2   After sterilisation, containers, if they are water cooled, must be cooled quickly down to temperatures of about 40ºC (104ºF); failure to do this may give rise to significant thermophilic flat-sour or blown container spoilage. Such spoilage may call into question the effectiveness of the thermal process and may, until a full bacteriological examination has been carried out, suggest a public health risk. Final temperatures of cooled containers are obtained by noting the equilibrium internal temperature of the contents.

17.2.3   The cooling operation, which immediately follows the sterilising stage, should be carried out carefully in order to avoid distortion and straining of the containers. Pressure cooling, in which air at slightly over the sterilising pressure is applied during the early stages of cooling, minimizes this strain. Reliable air pressure control instruments are essential but for normal requirements these need only be capable of controlling to about ± 0.1 kg/cm^2 (2 lbs/in^2).

17.2.4   The technique of the cooling operation will be governed by a number of factors. These include: the type or design of the cooker; the sterilising temperature; the size, diameter and substance of the food container or closure; the amount of product; internal container vacuum; and loading of the cooker. Seasonal temperature changes may also affect the length of the cooling time required.

17.2.5   Sudden release of the pressure within the cooker may result in permanent distortion at the double seam area of the can ends – a condition referred to as "peaking." Keeping the air pressure or water pressure unnecessarily high during cooling could result in permanent flattening of the can walls, a condition known as "panelling." These two problems are not usually encountered in can sizes of 75 mm (300) diameter or below. With aluminium containers and those with easy-open ends, special precautions during cooling are needed to avoid distortion and increased risk of leakage. Pressure cooling arrangements for this type of container are usually specified by the can manufacturer.

17.2.6   At the end of the heating operation (as cooling water is introduced into a retort loaded with hot containers), steam will condense, causing a loss in system pressure. To counteract this, compressed air is introduced at the time the steam is turned off, so as to maintain the pressure within the cooker at or slightly above the cooking pressure. With can sizes of 75 mm (300) diameter or below, the pressure within the water-filled cooker may be reduced to atmospheric and subsequent cooling of cans is continued for the required period at atmospheric pressure. With products in can sizes in excess of 75 mm (300) diameter, extended pressure cooling is required in which superimposed air pressure is maintained and carefully controlled within the cooker during cooling and gradually reduced stepwise to atmospheric during the latter stages of the cooling cycle. Cooling may be continued in the cooker until the can contents have reached the required temperature or, alternatively, when their internal pressures are such that no undue double seam straining will occur, cans may be removed from cookers and final cooling effected in a cooling canal. The cooling water feed into the canal should flow in the opposite direction to the cans and there must be sufficient flow to achieve cooling requirements. The water must be bacteriologically suitable and monitored in accordance with subsection 5.12.

17.3 WATER QUALITY FOR COOLING PURPOSES

17.3.1　A primary requirement for container cooling water is that it should be free from micro-organisms which might gain access to the cans during the cooling process. The total aerobic colony count should be less than 100 organisms per millilitre after incubation for five days at 20–22°C, and coliforms should not be detected in any sample.

17.3.2　Ideally, all cooling water should be obtained from a source of potable quality, i.e. either water authority mains or an uncontaminated well or bore-hole. The microbiological quality of the water from such a source or supply should be monitored by analysis, but vigilance is necessary, particularly in periods of climatic change, as significant variations in water quality may occur. It should also be noted that even potable water may contain bacteria in numbers which make it unsuitable for container cooling purposes. It is considered normal that water used for container cooling will have received a disinfection treatment, usually chlorination. If this is not undertaken it is important that the processor is able to demonstrate the consistent satisfactory microbial quality of the cooling water used.

17.3.3   Frequency of testing of cooling water for microbiological quality should be in relation to the confidence in the consistency of water quality. In general, it is recommended that the total aerobic plate count should be checked at least weekly and coliforms tested for monthly. Any significant variation from the established limits should be investigated and the level of sampling increased accordingly. The presence of coliforms in any sample indicates the need for immediate investigation.

17.3.4   WHEN COOLING WATER IS RECIRCULATED, there will be a tendency for the microbial population within the water to multiply and consequently A DISINFECTION TREATMENT IS NECESSARY. It is also greatly preferable that the treatment provides a residual disinfecting property to the water at the point of use.

17.3.5   When water is contaminated with a high level of organic impurity, either at source or as a result of continual recirculation, it is necessary to provide appropriate treatment such as filtration; and in a circulating system, it may be necessary to drain and refill the system periodically in order to remove gross debris. Build-up of impurities within the water may have a considerable effect on the efficiency of the disinfection process.

17.3.6   Be aware that non-potable water may be circulated within the premises for such purposes as steam raising or fire control. It should be carried in completely separate lines and these should be distinguished by distinctive colour coding. No cross connections with pipes carrying potable water should be permitted without break tanks being fitted.

17.4   CHLORINATION OF COOLING WATER

17.4.1   Chlorination is the most commonly used disinfection system for water, though other effective treatment systems may be used.

17.4.2   Chlorine gas or chlorine solution may be injected directly into the water, or it may be added as a solution of sodium or calcium hypochlorite. The amount of chlorine required will be affected by the organic load in the water.

17.4.3   For effective disinfection, after addition of sufficient chlorine the water must be thoroughly mixed and then held for not less than 20 minutes. The holding period is known as the contact time and the required contact time will be related to the individual chlorination system. Holding tanks should be so constructed as to ensure this minimum contact time, even at periods of maximum demand. The adequacy of the chlorination treatment is indicated by the presence of residual "free chlorine" in the water at the end of the contact time. Residual "free chlorine" should be measured with sufficient frequency to ensure that the chlorination system is operating as intended. Particular attention should be given while the correct chlorine treatment is being established. It is most important that residual "free chlorine" measurements are made correctly and that the basis of the chlorination process is understood. Reference should be made to the chemical suppliers' manual and to Campden Technical Manual No. 1, "Post-process Sanitation in Canneries". Can cooling water after use should still contain detectable amounts of residual "free chlorine". This should be measured after leaving retorts or continuous sterilisers at least four times a day. Automatic alarms may be used to indicate low levels of chlorine in cooling water.

17.5   ALTERNATIVE TREATMENTS TO CHLORINE

There are other alternatives to chlorination. Chlorine dioxide and bromine compounds may be used but there are hazards associated with their use, and detailed advice must be sought from the system manufacturers before installation.

17.5.1   **Chlorine Dioxide**
This sterilant has a number of advantages over chlorine when used as a disinfectant in cooling water systems.

a. Chlorine dioxide does not react with organic soils as chlorine does; therefore, the effectiveness is not influenced by fluctuating soil levels. (Similarly, problematical chloramines are not formed.)
b. Kill rate is not influenced by high pH as is the case with chlorine. Attempts in canneries to reduce water discharge rates may result, in certain locations, in substantial increases in water pH.

c. Rapid kill rates are possible (for example, 20 minute contact at 2 ppm ClO_2 reliably yields "sterile" water at pH 8).

7.5.2 **Bromine Compounds**
These slowly release bromine into the water and are much more active than chlorine at alkaline pHs.

17.6 POST-THERMAL PROCESS CAN CLEANING

Post-thermal process can cleaning is best regarded as a hazardous operation and will always carry some risk of post-process contamination. It cannot therefore be recommended as a routine procedure. If undertaken in an emergency, it is essential that the washing equipment is clean and the wash water controlled at a temperature above 85°C.

17.7 POST-THERMAL PROCESS CAN HANDLING

17.7.1 It has been well established that a small proportion of correctly made and seamed cans may be affected by minute temporary leaks (microleakage) during cooling and for as long as the cans remain externally wet. Moisture on the can bodies and covers provides a source of and a transport medium for micro-organisms from cooling water and conveying and handling equipment. The risk of microbial re-infection is increased by poor seam quality or by physical abuse of cans during conveying and handling.

Glass jar closures may be similarly affected.

17.7.2 To control microleakage, it is necessary to ensure that:
a. Wet containers are not manually handled;
b. Containers are dried as quickly as possible;
c. Conveying and handling surfaces are routinely cleaned and disinfected in order to reduce microbial contamination to a prescribed level; and
d. Conveying and handling equipment is designed, installed, operated and maintained so as to cause minimum physical abuse to the containers.

17.7.3 Unlike cans, correctly formed heat seals are not susceptible to temporary leaks. However, leakage may occur through defective seals and perforations in flexible or semi-rigid materials. Therefore, the requirements for drying containers, cleaning and disinfecting equipment surfaces and minimizing container abuse apply also to containers closed with seals.

17.8   RETORT BASKET UNLOADING

Manual unloading of wet containers presents a risk of contamination from food poisoning organisms, which may be transferred from the operator's hands into the container by microleakage. Containers must not be manually handled while still warm and wet, either from retort baskets or from runway systems. This can normally be achieved by ensuring that retort baskets, on removal from the retort, are as fully inclined as possible to allow time for water to drain from the countersink areas. Afterwards, cans should be allowed to cool and dry in the baskets before any attempt is made to remove them by hand.

17.9   CONTAINER DRYING EQUIPMENT

17.9.1   Driers will not remove all cooling water residues from can bodies but they will reduce significantly the time cans remain wet. For cylindrical containers, this partial drying can be effected by absorption methods, or for all containers by the use of high velocity air jets. It is essential that, whatever means are used for drying the cans, the equipment is situated in the line as soon as is practicable after the can cooling stage. Can drying units should not cause can damage and should be capable of ready access for routine sanitation. Cans should be visibly dry on leaving the drying unit but a simple test, involving the blotting of cans with pre-weighed filter paper followed by re-weighing, will give an indication of drier performance and allow suitable standards to be set.

17.9.2   To speed up the operation of the drying of retort processed cans, a method involving dipping the filled retort baskets in a tank of suitable wetting agent can be used. However, the dip time should not exceed 15 secs and the temperature of the wetting agent must not be less than 77ºC (170ºF). The wetting agent should be changed regularly.

17.10   POST-THERMAL PROCESS EQUIPMENT SANITATION

17.10.1 Inspection of can conveying equipment will reveal surfaces which are wet or likely to become wet during production periods. High numbers of bacteria may be expected on non-disinfected wet surfaces of can conveying equipment and, consequently, there will be a considerable risk of microleakage as cans pass through the line. Strict attention to plant hygiene is necessary in order to avoid this problem.

17.10.2　It is recommended, therefore, that all wet surfaces which are in contact with processed cans should be cleaned efficiently at least once every 24 hours. Any clean-down procedure, whether carried out before or after production, should be thoroughly evaluated before being adopted as a routine procedure.

17.10.3　It may be found necessary to use a combination of physical and chemical means of sanitation in order to achieve a recommended residual bacterial level not in excess of 500 organisms per 26 cm^2 (4 sq ins). The basic cleaning requirements to achieve this standard will vary according to the nature and complexity of the can conveying equipment involved.

17.10.4　In certain cases, it may also prove necessary to superimpose a continuous or semicontinuous method of disinfecting the can conveying system to operate during actual 7.9 production periods, in order to reduce bacterial contamination to less than 500 organisms 7.9.1 per 26 cm^2 (4 in^2). Manual or mechanical spraying with an appropriate disinfectant is suitable for this purpose.

17.10.5　Materials used as cleansing or disinfecting agents must be effective, non-corrosive and non-tainting. Choice of these materials, therefore, should be based on a thorough knowledge and understanding of their general properties and on their suitability for the particular application. Cresols, phenolics and some other types of disinfectants are suitable for post-process sanitation measures.

17.10.6　Certain areas of some continuous cookers and hydrostatic cookers may constitute a continual source of high bacterial infection unless stringent measures are taken to clean and disinfect them regularly and efficiently as part of the daily post-process hygiene routine.

17.10.7　Microbiological monitoring is necessary to establish a suitable hygiene schedule initially. It should also be undertaken as a regular surveillance procedure on the can handling system. Bacteriological swab tests offer a reliable means of assessing the efficiency of end-production clean-down procedures and other hygienic measures carried out on can conveying equipment.

17.10.8 All personnel in post-process areas should be fully instructed in the importance of personal hygiene and habits in relation to post-process can handling (see subsections 6.5 and 6.10). Operatives should be fully aware of their responsibility towards ensuring that can handling equipment is correctly and adequately cleaned and disinfected. Under no circumstances should manually handled cans be returned to the process line.

17.10.9 Post-process contamination and recommendations for cleaning and disinfection techniques are detailed in CFDRA Technical Manual No. 1. "Post-process Sanitation in Canneries". In addition, details of the hygienic design of post-process systems are described in Technical Manual No. 8.

17.11 LABELLING

It is good practice that cans and other containers for heat sterilised foods are labelled dry. Occasionally cans that are not completely dry are passed through labelling machines. This situation may arise because water is thrown from the seam areas by a change of speed or direction, or by condensation on overcooled cans, or the intermittent labelling of bright cans stored in cold conditions. Under these conditions, some evidence will be seen, as production continues, of build-up of water on the can tracks through the labeller. Therefore, the need for an identical standard of hygiene, coupled with avoidance of can abuse, applies as much to labelling machines as to other can handling equipment.

The label should be applied using a non-corrosive adhesive.

17.12 CASING/SHRINKWRAPPING

17.12.1 In order to avoid corrosion or "stack burn" problems, containers should be completely dry and at a suitably cool temperature before casing or shrinkwrapping.

17.12.2   Automatic casing of containers is preferable to manual casing. The latter should only be attempted after establishing that containers are externally dry. Automatic casers should be included with container conveying equipment as needing regular, routine cleaning and disinfection treatment. They should be adjusted so as to minimize container abuse. This also applies to shrinkwrapping equipment.

17.13   CAN STORAGE

As a means of delaying external and internal corrosion effects, attempts should be made to avoid wide variations of temperature and humidity during storage. Ideally, warehouse daily temperatures should not vary by more than 5–8°C (9–15°F). The higher the storage temperature, the greater the rate of corrosion. The maintenance of adequate ventilation and constant storage temperature is important for the prevention of external rusting caused by condensation.

17.14   "PROCESS AUDIT" RECONCILIATION

17.14.1   The objective of the "process audit" reconciliation is to ensure safety of production from continuous sterilisers or batch retorts. In any reconciliation, the following minima should be achieved.

17.14.2   All thermal process information must be reconciled, i.e. confirm that actual values agree with the scheduled values for time, temperature and pressure. The reconciliation should be completed by the morning of the next working day at the latest. The product details, process time and temperature should be recorded on the process audit sheet.

17.14.3   For continuous processes, the operator's process log and the recorder chart of the steriliser should both be examined to ensure that no undetected process deviations had occurred during the production.

17.14.4   For batch retort processes, the number of crates filled from the total production quantity being audited must be recorded. The retort check log should be checked that it has the correct product details recorded.

17.14.5   The number of cycles and the number of crates from the retort operator's log should be recorded on the audit sheet. The vent time, process time, projected let-down time and actual let-down time should be checked (the last two should agree). Each log should be signed after checking.

17.14.6   The number of cycles on each retort chart should be recorded and each cycle's vent time, process time and process temperature checked. Time and temperatures on the process traces should be recorded, and the final number of processes for each product recorded on the audit sheet.

17.14.7   The crate identification indicators should be counted and recorded on the audit sheet. Each indicator should be checked to ensure that a heat sensitive device is attached and has changed colour.

17.14.8   On the audit sheet, the following figures should agree:

a. The number of crates loaded;
b. The number of crates processed;
c. The number of crates checked;
d. The number of crate indicators;
e. The number of processes (retort crates, including part crates);
f. The number of processes (operator's log).

17.14.9   If all the details are correct, then the process audit sheet can be signed, dated and authorised. If the details disagree, then a full investigation must be undertaken and resolved in a satisfactory way. In the absence of a satisfactory explanation, then the product must be isolated until the process effectiveness has been established.

18 INCUBATION TESTS

18.1   Introduction

18.2   Laboratory Incubation

18.3   Bulk Incubation

18.1   INTRODUCTION

18.1.1   Properly designed incubation tests may provide valuable additional information, but the results should never be used as the sole criterion of assessing the safety or acceptability of any sterilising process or the microbiological status of any quantity of production.

18.1.2   The type of incubation test used is a responsibility of the individual manufacturer but tests can be considered under two main headings.

18.2   LABORATORY INCUBATION

18.2.1   Incubation tests in the laboratory are normally relatively small scale tests, followed by destructive examination of the samples, and may be undertaken for a variety of reasons. Their statistical soundness varies considerably and, because of this, it is essential that the purpose of the test and its statistical basis are clearly understood before any decisions or predictions are made from the results.

18.2.2   The temperature of incubation and the subsequent methods of examination of the samples need to be chosen with great care in order to ensure that an incubation test is valid for the purpose intended (see also subsection 18.3.3). In particular, it should be remembered that a spoiled container will not necessarily "blow" or give any other external signs of spoilage. It is therefore necessary to open all incubated sample containers for appropriate examination.

18.2.3   It is not widely appreciated that, in order to detect spoilage with a reasonable degree of certainty, a very large number of containers would need to be subject to an incubation test, even when the level of potential spoilage in the lot is fairly high. For example, if the true defect rate is one can in a thousand, to be sure of detecting this rate in at least 95% of samples, each sample would need to consist of 3,000 cans; to be sure of detecting it in 99% of samples, each sample would need to consist of 4,500 cans.

18.2.4   To construct a suitable incubation scheme, it is necessary to decide what probability of acceptance of lots with a given proportion of defectives can be tolerated. It can then be determined from cumulative probability curves how many samples should be incubated in order to ensure that this tolerance level is not exceeded (BS 6001 – Sampling Procedures and Tables for Inspection by Attributes, 1991).

18.2.5   Once this decision has been made, the operating characteristics curve of the scheme can be drawn to show the behaviour of the incubation test under the various conditions likely to be encountered. The method for constructing operating characteristics curves may be found in many standard quality control texts (e.g. Juran, 1974).

18.3   BULK INCUBATION

18.3.1   Bulk incubation can only practicably be used to establish the incidence of spoilage producing "blown" containers. The construction of such a scheme will depend, as previously mentioned, upon the tolerances involved.

18.3.2   If the results of routine bulk incubation tests are plotted consecutively in chronological order on a control chart graph by variety, line and can size, any trend towards increasing spoilage levels may be seen and action can be taken to remedy the causes.

18.3.3   Experience has shown that it may take up to a week for all cans in a large stack to reach incubation temperature and, for this reason, it is advantageous to place the cans in the incubation room as soon as possible after retorting while they are still warm.

19 EMERGENCY PROCEDURES

19.1   PROCESS DEVIATIONS

There should be instructions, authorised by the Thermal Process Manager, which define the actions to be taken in the event of thermal processes which are observed to deviate from their scheduled conditions. If product has been overprocessed, the issue will normally be one of quality, but if underprocessed, the product will require to be further sterilised or destroyed.

19.1.1   Process Extension

If a process deviation is noted during the thermal process, it may be possible to extend the process to compensate for the deviations from scheduled conditions. The extension should be sufficient to attain the required lethality and should be in accordance with previously established procedures.

19.1.2   Reprocessing

Any reprocessing operations which are carried out should be performed under the supervision of the Thermal Process Manager, who will need to consider all the factors involved, including any changes to product characteristics which may have taken place which would need prior verification or simulation of the proposed reprocess.

N B A reprocess may need to be more severe than the originally scheduled process due to changes in the nature of the product which affect heat transfer.

19.1.3	Reprocessing, where appropriate, should take place within a short time of the failed process under the direction of the Thermal Process Manager.
19.1.4	No reprocesses should be carried out on batches which have been previously processed and subsequently exhibited microbial spoilage.

19.2 RECOVERY OF DISTRESSED BATCHES OF PRODUCT

The recovery of distressed batches of product should only be carried out by trained personnel under the direct supervision of person(s) having expert knowledge of heat processing and container technology, and in working conditions which provide the lighting, space and facilities necessary for the task.

The hazard analysis critical control points (HACCP) concept should be applied in the determination of the appropriate procedures and should include:

19.2.1	An assessment of the hazards associated with the adverse conditions which led to the product being suspect, and any residual hazards which would remain after the various recovery options under consideration had been completed.
19.2.2	Identification of the critical control points for the recovery operation, and the type and frequency of the control measures deemed necessary.
19.2.3	Guidance for the monitoring of the critical control points in order to provide assurance of the effectiveness of the procedure and for the maintenance of adequate records.

19.3 RECALL PROCEDURES

There should be a predetermined written plan, clearly understood by all concerned, for the recall of a product, or a batch or batches of product, known or suspected to be hazardous or otherwise unfit, or of wholesome but substandard product which the manufacturer wishes to recall.

19.3.1	A designated person, with appropriate named deputies, should be nominated to initiate and coordinate all recall

activities, and to be the point of any contact with the local Environmental Health authority and the Department of Health on recall matters.

19.3.2   The design of manufacturing records systems and distribution records systems, and the marking of outer cartons and of individual packs, should be such as to facilitate effective recall if necessary.

19.3.3   There should be a written recall procedure, and it should be capable of being put into operation at short notice, at any time, inside or outside of normal working hours. Specifically, this includes weekends and public holidays.

19.3.4   The recall procedure should be shown to be effective within a reasonable time of implementation, and at suitable intervals thereafter, by carrying out suitable testing of the procedures.

19.3.5   The recall procedure should be reviewed regularly to check whether there is need for revision (e.g. telephone list update).

19.3.6   Although a defect or a suspected defect leading to recall may have come to light in respect of a particular batch or batches or a particular period of production, urgent consideration should be given to whether other batches or periods may also have been affected (e.g. through use of a faulty material, or a plant or processing fault), and whether these should also be included in the recall.

19.3.7   The recall procedure should lay down precise methods for notifying and implementing a recall from all distributive channels and retailers where the affected product might be, as well as affected goods in transit, and of halting any further distribution of affected goods.

19.3.8   Notification of recall should include the following information:

a. Name, pack size and adequate description of the product;
b. Identifying marks of the batches concerned and their location;
c. The nature of the defect;
d. Action required, with an indication of the degree of urgency involved.

19.3.9   Recalled material should be quarantined in a safe location, pending decision as to appropriate treatment or disposal.

19.3.10   Up-to-date contact lists for suppliers, distributors, retailers, enforcement authorities, media, legal consultants and external laboratories should be established and maintained.

19.4   SABOTAGE

Regrettably, the possibility of real or threatened hazard arising from the actions of second or third parties must be faced, e.g. deliberate contamination or poisoning of product or ingredient.

19.4.1   The first intimation of a problem in this area could come from a whole variety of sources, e.g. consumer complaint, from an Environmental Health Officer, employees, or by telephone, post or personnel contact with any company location or any employee at any time. It is prudent in these situations that prior contact be established with the appropriate statutory food authorities, and that written procedures exist to notify the said authorities and the police as early as possible in case of such an incident.

19.4.2   It is essential that all personnel should be aware of company procedures to be followed in dealing with such threats, both within and outside of normal working hours, and that suitable arrangements exist for calling in key personnel out of hours in such an emergency. The extent to which any such emergency procedures may override normal lines of management should be explicitly stated.

19.4.3   The possibility of such sabotage and even site invasion may indicate a need for particular security precautions in vulnerable areas, e.g. locked rooms, use of seals, etc.

19.4.4   Faced with an emergency situation, the recall procedures described in subsection 19.3 apply, while the expertise of those involved in quality control, assurance and audit should be put at the disposal of the management responsible for handling the emergency.

20 FACTORY CLEANING PROCEDURES

20.1   General

20.2   Cleaning Operations

20.3   Sanitation of Post-process Can Handling Equipment

20.1   GENERAL

20.1.1   Cleaning of the plant and equipment form an important part of the overall sanitation of the factory. It can have a direct effect upon the quality and safety of the products produced.

20.1.2   Management should recognise and accept a direct responsibility for the maintenance of proper standards of cleaning.

20.1.3   Cleaning should be a properly planned and managed function. It is important that all cleaning procedures used in a factory have been properly evaluated and are written down in a cleaning schedule. This schedule should set out the frequency of the cleaning procedures throughout the factory and should specify material, equipment and methods needed for each item of processing equipment and each area of the premises. All operatives responsible for cleaning should be trained in the proper use of equipment and the proper methods of cleaning required in the plant as set out in the schedule.

20.2   CLEANING OPERATIONS

20.2.1   Cleaning consists of removal of soil (as defined in subsection 3.70), slime and micro-organisms by physical, chemical or mechanical means from surfaces. Cleaning may be followed by disinfection as defined in subsection 3.27.

20.2.2   The efficiency of a disinfecting process will depend upon the efficiency of the removal of soil prior to application of the disinfecting agent.

20.2.3   The methods adopted should be in accordance with, but not necessarily limited to, those recommended in standard texts on cleaning.

20.2.4   Frequency of cleaning should depend upon the type and use of equipment but, generally speaking, all processing equipment and areas should be cleaned at a frequency which prevents product deterioration, and most areas should be cleaned daily as soon as production ceases.

20.2.5   For extended production programmes, the frequency at which a line needs to be closed down temporarily for sanitation and housekeeping purposes should be established as part of the quality assurance programme, wherever possible supported by a line control, e.g. HACCP, study. During planned breaks, care should be taken to ensure that no exposed product or packaging material is contaminated during the cleaning operations.

20.2.6   The effectiveness of cleaning programmes should be regularly assessed and records retained. The type of assessment may vary depending upon the stage in the manufacturing operation; for example, simple visual and tactile examinations would generally be adequate for food preparation equipment, whereas bacteriological tests are necessary to evaluate the status of container handling systems.

20.3   SANITATION OF POST-PROCESS CONTAINER HANDLING EQUIPMENT

20.3.1   To avoid post-processing contamination, all post-process container handling and contact surfaces should ideally be kept dry. Special attention needs to be given to the cleaning and disinfection of post-process container handling equipment.

20.3.2   For details of the post-process contamination and the cleaning and disinfection techniques recommended, see subsection 17.10 and Technical Manual No. 1, "Post-process Sanitation in Canneries," issued by the Campden Food and Drink Research Association.

21 ASEPTIC PRODUCTION TECHNOLOGY

21.1   Introduction
21.2   Products
21.3   Critical Features of Plant Design
21.4   Selection of Sterilisation Systems
21.5   Pre-production Sterilisation of Equipment
21.6   The Evaluation of a Heat Process for Food
21.7   Aseptic Packaging and Filling
21.8   Cleaning in Place
21.9   Training
21.10   Control and Instrumentation

12.1   INTRODUCTION

Aseptic technology is complex in that sterilisation of the product is carried out as a separate operation from packaging. For this reason, unless production systems are designed, constructed and operated on sound principles, aseptic technology may produce risks of unsterility or potential harm to the consumer. Three key points are set out below:

a. Everyday control depends upon sophisticated monitoring and control systems to assure that each step of the manufacturing scheduled process is achieved.
b. The same standards which are applied to in-container processed foods apply to aseptically processed foods in respect of public health and safety.
c. Because the foodstream is processed separately, the scheduled heat process is the same, irrespective of packaging size.

Further information may be found in documents produced by the Campden Food and Drink Research Association: (1) Technical Manual No. 11, "Good Manufacturing Practice Guidelines for the Processing and Aseptic Packaging of Low-Acid Foods: Part 1 – Principles of

Design, Installation and Commissioning; Part 2a – Test Methods in Design and Commissioning; Part 2b – Test Methods in Production"; (2) Technical Manual No. 24, "The Microbiological Aspects of Commissioning and Operating Aseptic Production Processes".

21.2   PRODUCTS

The choice of plant and scheduled process to achieve commercial sterility will depend on the nature of the food to be processed.

21.2.1   The scheduled process required depends on the acidity of the product, just as with in-container processing. Thus, any component which has a pH value above 4.5 after processing will define the whole system as low acid for which a minimum botulinum heat process will be necessary.

21.2.2   The flow characteristics of the food in a specific system will affect the residence time distribution in a holding section, and hence the scheduled heat process. It is the slowest heating part or particle which is important – not the average.

21.2.3   The rheological properties of the food should be known since its behaviour under different conditions of shear or temperature may affect pump efficiency, the consequent flowrate, and hence the scheduled heat process.

21.2.4   The electrical properties of the material will be critical in processes using electrical resistance heating.

21.2.5   The size, concentration and composition of any particulate material may affect the overall heat transfer characteristics of the product and could cause problems due to blocking or channelling.

21.2.6   Any tendency of a food to cause fouling or burn-on in the heat exchange system will be important because of the effects on heat transfer, blocking, and the reduced residence time in the holding section.

21.3   CRITICAL FEATURES OF PLANT DESIGN

21.3.1   Product and process data are required before the suitability of plant design can be assessed.

21.3.2   It may be dangerous to the consumer to process product on a plant which is unsuitable.

21.3.3   Plant should be constructed for operator safety and in accordance with sound aseptic practice (e.g. see CFDRA Technical Manual No. 11, "Good Manufacturing Practice Guidelines for the Processing and Aseptic Packaging of Low-Acid Foods".

21.3.4   Any surface in contact with commercially sterile food should be non-porous and free from cracks through which micro-organisms could penetrate.

21.3.5   Devices and joints should be free from crevices where product may lodge and may become infected if not removed by CIP or adequately sterilised.

21.3.6   "Dead ends" in pipework should be avoided wherever possible. If they do exist, they should be positioned to ensure that turbulent flow flushes them.

21.3.7   Process design dictates the type of pump needed to transport product. In a continuous plant, the pump must be able to deliver water and product against the system pressure at a uniform and controlled rate, since this is the time base of the thermal process. Selection of the wrong pump is one of the major factors affecting stability of operation.

21.3.8   Pumps, such as back-pressure or booster pumps, in contact with commercially sterile food should be of aseptic design. These pumps must be constructed from materials which will withstand the temperatures of heat sterilisation. Shaft seals should be protected against microbial ingress by, for example, steam vented to atmosphere. All internal surfaces must be easy to clean and heat sterilised.

Valves should be capable of thorough in-place cleaning (CIP) and being rendered commercially sterile during the normal course of plant sterilisation, and afterwards held in this state throughout a production run. This applies to all surfaces that contact food directly in either the open or closed position. Diaphragm-type valves should be routinely serviced and inspected. When plant is not operational, they should be left open.

21.3.9   CIP pumps should be able to deliver sufficient flowrate velocity to adequately clean the fouled surfaces.

21.3.10   The design of the holding section is central to the function of heat sterilisation since its performance determines the time and temperature components of the scheduled heat process.

The process temperature should be measured at the inlet and outlet to the holding section.

When a tube is used as a holding section, it may be lagged but should not be heated.

The sizing of holding tubes is designed to accompany a scheduled process. Pipe diameter and length must be checked against specified design to ensure correct residence time.

Holding tubes should be documented and indelibly marked to identify individual sections and to ensure correct assembly.

Many foods are non-Newtonian fluids with high apparent viscosities exhibiting laminar flow. Additional allowance is required for conduction or equilibration of heat within particles.

Holding sections should be designed to prevent entrapment of air. Where the section comprises a tube, an elevation of not less than 20 mm per 1000 mm of pipe should be made.

21.3.11   Steam in contact with the product or with product contact surfaces must be of culinary grade (see subsection 5.14)

Connections of steel service pipework to stainless steel process pipework should be via a steam strainer or other fine-mesh filter so as to prevent product contamination by rust particles.

21.3.12   Since flash evaporative cooling chamber pressures will be lower than ambient, the system should be protected from sources of infection caused by ingress of air. Locations to be guarded against include:

a. Flange seals;

b. Vacuum pump seals;

c. Product pump seals (where the product is pumped at a higher pressure).

The apparatus may be installed in such a way that enables it and any associated pumps, condensers, coolers, valves or pipework to be simultaneously in-line sterilised. Correct operation depends on the presence of a fixed hydraulic head in the basal outlet of the evaporator.

21.3.13   The microbiological integrity of heat exchangers will depend on all gaskets and seals maintaining an effective barrier between sterile and non-sterile liquids. This is aided in plate coolers by double separation between gaskets. A sustained pressure differential of at least 50 kPa on the product side over the coolant should be maintained. Corrosion of plates and tubes, especially by chlorine in cooling water or chloride in product, may result in pinholing with loss of product sterility.

21.3.14   Scraped surface heat exchangers are vulnerable because of possible imperfections of the seals to the rotating shaft bearings. These should be of a type protected by steam, which is vented to atmosphere during pre-sterilisation, or by a flow of sterile water when running. When sterile water is used, it is usually condensate. Such a system requires careful design to be effective.

21.3.15   It is important that, in batch processing plant, the design shall enable in situ product cooling to be carried out in such a way that contaminant micro-organisms cannot be drawn into the reactor by the formation of partial vacua. This will depend on bearing seal integrity, backed by the application of steam or sterile condensate lubrication as necessary, together with a capability for generating an internal overpressure by pumping sterile air or other gas into the reactor to maintain positive pressure.

21.3.16   When equipment, such as a product homogeniser, is installed for aseptic operations, then a means should exist to sterilise it before production by use of steam or pressurised hot water. Any rotary or reciprocating drive shafts should be protected against microbiological ingress by high pressure steam as a preferred medium, and steam barriers should protect the shafts of pressure adjustment devices.

21.3.17   Aseptic tanks are used in order to separate, if needed, the operations of sterile the product manufacture and aseptic packaging. The tank must be a pressure vessel designed to withstand full vacuum, and capable of being sterilised with saturated culinary steam in the empty state under pressure, with adequate steam penetration into all ancillary pipes, connections, valves, pressure gauges and sample ports. Means must exist for draining off condensate. All inlets and outlets must be capable of double isolation, preferably with a provision of a steam barrier being established after the empty vessel sterilisation cycle. It must be possible to attain a state of commercial sterility within the tank and, for this purpose, heat sensor probes must be attached. If necessary, these should be placed taking account of any "cold spots."

The tank must be provided with means by which a positive internal pressure may be maintained and controlled within the headspace by admission and venting of sterile inert gas or air. This air should be freed from viable organisms by passage through a steam sterilised absolute filter and/or by incineration.

If an impeller is included for product bulk mixing, then the shaft seals should be protected against microbial ingress by, for example, steam vented to atmosphere.

21.3.18   Filling lines and fillers must undergo pre-production decontamination, either by simultaneous or separate heat sterilisation, usually by steam from the aseptic tank. Cold spots must be identified during the design stage and take into account the time/temperature requirement needed.

21.3.19   Pre-production decontamination must reduce the microbiological population such that any residual will not affect the commercial sterility of the product. The lethality of this process must not be less than that applied in the scheduled heat process of the food itself.

21.4   SELECTION OF STERILISATION SYSTEMS

21.4.1   The choice of plant and scheduled process to achieve commercial sterility will depend on the nature of the foodstuff to be processed.

21.4.2   The plant consists of the pre-processing, heating, holding and cooling sections, terminating at the aseptic tank or filler.

21.4.3   Processing may be batch, continuous or hybrid. Streams of sterile product may be combined at various stages, including the aseptic tank or filler.

21.4.4   Heating and cooling may be by direct or indirect means.

21.4.5   Novel direct heating systems may use electrical energy instead of steam. The energy and heat transfer mechanisms are different and should be understood.

21.5   PRE-PRODUCTION STERILISATION OF EQUIPMENT

Pre-production sterilisation of food contact surfaces by heat must at least provide a microbiological lethality similar to that required by the food or drink. The critical control point is that showing the worst-case condition, such as the slowest heating plant item. Following this process, suitable precautions must be taken to avoid re-infection, especially during cooling.

21.6   THE EVALUATION OF A HEAT PROCESS FOR FOOD

The principles of the thermal process set forth in sections 10, 11 and 12 for in-pack sterilisation apply.

The evaluation of a heat process for food in continuous-flow processing must take into account the minimum residence time of a fluid element or particle in the holding section and the lowest temperature achieved within the food. In batch processes, the time/temperature relationship is also critical.

Whether the process is pasteurisation, as taken in the broader sense, or full sterilisation depends on the nature of the food. In both cases the objective is to achieve commercial sterility and extreme caution must be advised for low acid foods. Experimental evidence and theoretical considerations will provide the rationale for increased process lethality to ensure product safety. Within emerging food processing technologies such as electrical resistance heating, electrical properties of foods must be considered also.

21.7   ASEPTIC PACKAGING AND FILLING

21.7.1   There are many varieties of primary packaging, which may be either rigid, semi-rigid or flexible. The requirement of an aseptically filled package is that it will be commercially sterile. This requires that the food or drink must be commercially sterile and that the food contact surfaces of the packaging must be sufficiently decontaminated from microorganisms. The filling operation must preclude re-infection and the aseptic seal of the primary package must make an effective barrier against entry by micro-organisms.

21.7.2   Maintenance of pack integrity throughout storage and distribution will depend on knowledge of the factors critical to the performance of the packs, and their control. The chemical nature of a food must not cause it to react so as to result in pack weakness or damage. It must be remembered that in many cases only the inner surface of a package is decontaminated. Any perforation of this inner surface may expose the product to micro-organisms within the laminate. Fill weight control is important, either to confer sufficient hydraulic resistance or to avoid over-stressing the package and its seals.

21.7.3   Microbiological decontamination of packaging must be defined. A number of methods are used and in each case factors which are critical must be understood and controlled. Chemicals, especially hydrogen peroxide, are frequently applied. Factors critical to their effectiveness can include type, concentration, method and amount of application, coverage, temperature and contact time. The packaging materials must be able to withstand the method used.

21.7.4   It is essential that the establishment and maintenance of asepsis in a filling zone is achieved through a method appropriate to each system. The assurance of asepsis can only be gained through study of the critical control points of each machine. This must include the ability to clean and disinfect actual or potential product contact surfaces.

21.7.5 Where potentially hazardous chemicals are employed, either in plant pre-production sterilisation of non-food contact surfaces or in packaging decontamination, it must be established that residual levels do not comprise a hazard to the consumer.

21.8 CLEANING IN PLACE

Although the most severe fouling effects may often occur in heat exchangers, it is important to consider food contact surfaces throughout the whole installation. Some plant items, e.g. pumps, homogenisers and tanks, may require special cleaning methods. Operational procedures should avoid the possibilities for residues of cleaning compounds to contaminate foods.

21.9 TRAINING

Plant operators must have the ability to meet the requirements of sophisticated machinery and short response times. Key personnel should be trained prior to commissioning and must understand not only their own jobs but how they relate to the total operation.

21.10 CONTROL AND INSTRUMENTATION

21.10.1 The control system should ensure that the correct product sterilisation temperature and holding time are achieved, as defined by the scheduled thermal process.

The control system should be designed to "fail safe" in the event of a malfunction or loss of services. Understerilised food should be routed out of the system and not transferred to the aseptic storage or filling section.

The complexity of control and data logging systems depends on the configuration of the heat sterilisation plant and to the extent by which the controls are activated by automatic, manual or computerised means.

Factors such as temperature, pressure and rate of flow are critical to the process and therefore must be measured, controlled, indicated and recorded. An event log is invaluable. Records should be retained for a minimum of three years.

It should be taken into account that during plant commissioning or at other maintenance periods, engineers will set up or deactivate computerised and electromechanical safety interlocks. Plant operators and other unauthorised personnel should never be encouraged or permitted to alter, change or override these.

21.10.2 The siting of temperature sensors within continuous processing systems must be carefully considered. The following locations are essential:

a. Pre-heat of food;
b. Exit point of holding section;
c. Aseptic steam barriers.

The following locations provide important information and should be measured if possible:

d. Entry to holding section;
e. First stage of cooling;
f. Exit point to aseptic tank or packaging machine;
g. Aseptic tank sterile filter bleeds;
h. Steam barriers on the seals of rotary and reciprocating equipment such as homogeniser pistons, scraped surface heat exchangers and back-pressure pumps.

The way in which temperature probes are installed should not put the asepsis of a system at risk. They should properly sense representative streams and should not be subject to interference in output by adjacent equipment.

When particulate matter is included in the food, then probes must be designed and installed in a way which will not cause blockage.

21.10.3 In continuous systems, the temperature probe dimensions must be compatible with the diameter of the pipe. Mass should be minimized to reduce thermal lag and conduction errors.

The total tolerance on temperature measurement accuracy shall not exceed + 0.5°C when measuring temperatures in the range of 0–150°C, with measuring element in an environment between 20–45°C.

The maximum time allowed for a response to a change of temperature from 115 to 140°C shall be 1 minute, i.e. the indicator shall display 140 ± 0.5°C after 1 minute.

Alternatively, a response of 63% of the change from 115 to 140°C in not more than 20 seconds is acceptable, i.e. the thermometer must reach 121°C.

21.10.4   The product temperature at the entrance to the holding section should be maintained to within ± 1°C of the controller set point. The system must accommodate some variations in temperature and flow of the feed, and also, for example, changing heat transfer characteristics due to scale deposition on heat exchange surfaces.

21.10.5   The product flowrate is critical in continuous systems since it will determine the average residence time at the selected temperature in the holding section. The flowrate should be controlled and continuously recorded.

21.10.6   A pressure transducer should be located on or immediately adjacent to the holding section to monitor this parameter. Transducers of the membrane type should have good integrity since leakage could harbour a pocket of infection. The pressure of the product steam within the continuous system should be such that product will not boil even if it reaches the temperature of the heating medium.

21.10.7   The process of heat exchange is characteristically slow in a batch system.

A system for monitoring and recording pressure and temperature within the process vessel should be provided.

When temperature is measured by a platinum resistance retort thermometer (PRRT), it should conform to Campden Technical Bulletin No. 61 (1987).

22 REFERENCES

British Standards Institution (1991). BS 6001 Sampling Procedures and Tables for Inspection by Attributes. Part 1: Specification for Sampling Plans Indexed by Acceptable Quality Level (AQL) for Lot-by-lot Inspection.

CFDRA (1977). Guidelines for the Establishment of Scheduled Heat Processes for Low-acid Foods. Technical Manual No. 3, Campden Food and Drink Research Association, Chipping Campden, Glos.

CFDRA (1986). Good Manufacturing Practice Guidelines for the Processing and Aseptic Packaging of Low-acid Foods. Technical Manual No. 11 (Part 1: Principles of Design, Installation and Commissioning (1986); Part 2A: Test Methods in Design and Commissioning; Part 2B: Test Methods in Production (1987)), Campden Food and Drink Research Association, Chipping Campden, Glos.

CFDRA (1989). The Microbiological Aspects of Commissioning and Operating Aseptic Production Processes. Technical Manual No. 24, Campden Food and Drink Research Association, Chipping Campden, Glos.

Department of Health and Social Security and the Ministry of Agriculture, Fisheries and Food (1970). Food Hygiene (General) Regulations 1970 (SI 1970 No. 1172 as amended SI 1990 No. 1431 and SI 1991 No. 1343).

Department of Health and Social Security, Ministry of Agriculture Fisheries and Food, Scottish Home and Health Department, Department of Health and Social Services Northern Ireland, Welsh Office (1981). Food Hygiene Codes of Practice. 10: The Canning of Low Acid Foods. HMSO, London.

Food Safety Act 1990.

Heinz (1991). Quality Audits. Chapter 2c in "Principles and Practices for the Safe Processing of Foods", D.A. and N.F. Shapton (Eds.), Butterworth-Heineman, London.

HSE (1987). Steam Boiler Blowdown Systems. Health & Safety Executive Guidance Note PM60.

HSE (1989). Automatically Controlled Steam and Hot Water Boilers. Health & Safety Executive Guidance Note PM5.

ISO 8402 (1986). Quality – Vocabulary. Available as British Standard 4788, Quality Vocabulary Part 1:1987, International Terms. Also available in BSI Handbook 22, British Standards Institution, London.

Juran, J.M. (Ed.) (1974). Quality Control Handbook. 3rd Edition, McGraw Hill Books Inc., London.

MPMA (1989). Recommended Industry Specifications for Open Top Processed Food Cans. Metal Packaging Manufacturers Association, Slough, UK.

Richardson, P.S. and Bown, G. (1987). A Standard for Platinum Resistance Thermometers for Use on Food Industry Sterilisers and Pasteurisers. Technical Bulletin No. 61, Campden Food and Drink Research Association, Chipping Campden. Glos.

Sawyer, L.B. (1983). The Practice of Modern Internal Auditing. 2nd Edition, Institute of Internal Auditors, Altamonte Springs, Florida 32701, USA.

Shapton, D.A. (1986). Canned and Bottled Food Products. Chapter 7 in "Quality Control in the Food Industry", Vol. 3, 2nd Edition, Ed. S.M. Herschdoerfer, Academic Press. London.

SI 1602(1960). Food Hygiene (Docks, Carriers etc) Regulations 1960 as amended.

SI 791 (1966). Food Hygiene (Markets, Stalls and Delivery Vehicles) Regulations 1966 as amended.

SI 1546(1988). Public Health (Infectious Diseases) Regulations 1988 as amended.

WHO (1971). International Standards for Drinking Water. 3rd Edition, World Health Organisation, Geneva.

23 BIBILIOGRAPHY

There are a large number of publications which are helpful to the processor of heat sterilised foods. The following may be found useful in conjunction with this code of practice.

British Standards Institution

BS 1133: Packaging Code. Section 10.1, Tins and Cans.

Campden Food & Drink Research Association

Post Process Sanitation in Canneries. Technical Manual No. 1 (1968).

Canning Retorts and Their Operation. Technical Manual No. 2 (1975).

Guidelines on Good Manufacturing Practice for Sterilisable Flexible Packaging Operations for Low-acid Foods. Technical Manual No. 4 (1978).

Process Control in Hydrostatic Cookers. Technical Manual No. 5. Part 1: Validification of Cooker Operating Conditions (1981). Part 2: Factors Affecting Heat Penetration Rates (1984). Part 3: Guidelines on Emergency Procedures (1984).

The Heat Processing of Uncured Canned Meat Products. Technical Manual No. 6 (1984).

Hygienic Design of Food Processing Equipment. Technical Manual No. 7 (1983) (reprinted 1992).

Hygienic Design of Post-process Can Handling Equipment. Technical Manual No. 8 (1985).

Examination of Suspect Spoiled Cans. Technical Manual No. 9 (1985).

Visual Can Defects. Technical Manual No. 10 (1984).

Guidelines for the Establishment of Procedures for the Inspection of Canneries. Technical Manual No. 12 (1986).

Campden Food & Drink Research Association (cont)

Examination of Suspect Spoiled Cans and Aseptically Filled Containers. Technical Manual No. 18 (1987).

Process Control in Reel and Spiral Cooker/Coolers. Good Manufacturing Practice Guidelines. Technical Manual No. 26. Part 1: The Operation of Continuous Cooker/Coolers (1990).

The Processing of Canned Fruit and Vegetables. Technical Manual No. 29 (revised edition 1980).

The Shelf Stable Packaging of Thermally Processed Foods in Semi-rigid Plastic Barrier Containers. A Guideline to GMP. Technical Manual No. 31 (1991).

HACCP: a Practical Guide. Technical Manual No. 38 (1992).

Department of the Environment

Water Supply (Water Quality) Regulations 1989 (SI 1989 No. 1147) as amended 1991 (Sl 1991 No. 1837).

Private Water Supplies Regulations 1991 (SI 1991 No. 2790).

Department of the Environment, Department of Health & Social Security, Public Health Laboratory Service

Reports on Public Health & Medical Subjects. No. 71: The Bacteriological Examination of Drinking Water Supplies. HMSO (1982).

European Economic Community

Council Directive relating to the quality of water intended for human consumption (80/778/EEC).

Food and Drug Administration

Thermally Processing Low-acid Foods Packaged in Hermetically Sealed Containers. FDA Regulations 21, Part 113 (1983).

Food Processors Institute

Canned Foods, Principles of Thermal Process Control, Acidification and Container Closure Evaluation. 5th Edition (1988).

Heinz

Principles and Practices for the Safe Processing of Foods. D.A. and N.F. Shapton (Eds.), Butterworth-Heineman, London (1991).

Hersom, A.C. and Hulland, E.D.

Canned Foods: An Introduction to Their Microbiology. Seventh Edition, Churchill Livingstone Edinburgh London & New York (1980).

Institute of Food Science & Technology

Food and Drink. Good Manufacturing Practice: A Guide to its Responsible Management. Third Edition (1991).

Metal Box plc (now CMB Foodcan)

Double Seam Manual (1978); Addendum (1984).

National Food Processors Association

Thermal Processes for Low Acid Canned Foods in Metal Containers. NFPA Bulletin 26-L. 12th Edition.

24 ACKNOWLEDGEMENTS

Ambrosia Creamery   Mrs A. Olivant

Anglia Canners Ltd   Mr I. Tilley

Asda Stores Ltd   Mr G. Austin, Mr D. Brackston

Brooke Bonds Foods Ltd   Mr K.G. Anderson

Campden Food & Drink   Mr L. Bratt, Mr R.H. Thorpe.
  Research Association   Mr P.S. Richardson,
                          Mr D. Rose, Mr K.L. Brown

CMB Technology   Mr P. Bean

Co-operative Wholesale Society Ltd   Mr M. Oakes

Department of Health   Mr E.W. Kingcott

Gerber Foods International   Mr D. Jean

H.J. Heinz Co Ltd   Mr D. Shapton. Dr K. Dow,
                       Mr M.P. Jones

Hillsdown Ltd   Mr G. Burrows

HP Foods Ltd   Mr M. Page

John West Foods Ltd   Mr R.J. Footitt

Londreco   Mr M. Treagus, Mr M. Cresswell

MAFF Food Science Division   Dr R. Mitchell

Marks & Spencer plc   Mr T. Sawers

Master Foods Ltd   Mr S. Tearle

Nestlé Co Ltd   Ms A. Kennedy

Pedigree Petfoods   Mr D. Ritchie

Princes Foods Ltd   Mr D. Atherton

J. Sainsbury plc   Mr A. Leighton

Spirax-Sarco Ltd   Mr R. Glassonbury

Unilever Research   Mr M. Brown

Printed in the United Kingdom for HMSO.
Dd.297830, C25, 6/94, 3400, 5673, 287827.